# Unraveling Creation and Evolution Through Science and the Bible

# Unraveling Creation and Evolution Through Science and the Bible

## Exploring Mysteries, Facts, Theories, and Myths

**PAUL BRYAN**

RESOURCE *Publications* • Eugene, Oregon

UNRAVELING CREATION AND EVOLUTION THROUGH SCIENCE AND THE BIBLE
Exploring Mysteries, Facts, Theories, and Myths

Copyright © 2025 Paul Bryan. All rights reserved. Except for brief quotations in critical publications or reviews, no part of this book may be reproduced in any manner without prior written permission from the publisher. Write: Permissions, Wipf and Stock Publishers, 199 W. 8th Ave., Suite 3, Eugene, OR 97401.

Resource Publications
An Imprint of Wipf and Stock Publishers
199 W. 8th Ave., Suite 3
Eugene, OR 97401

www.wipfandstock.com

PAPERBACK ISBN: 979-8-3852-4716-5
HARDCOVER ISBN: 979-8-3852-4717-2
EBOOK ISBN: 979-8-3852-4718-9

05/15/25

Scripture quotations marked (NLT) are taken from the Holy Bible, New Living Translation, copyright ©1996, 2004, 2015 by Tyndale House Foundation. Used by permission of Tyndale House Publishers, Carol Stream, Illinois 60188. All rights reserved.

Scripture quotations by permission. Quotations designated (NET) are from the NET Bible® copyright © 1996, 2019 by Biblical Studies Press, L.L.C. http://netBible.com. All rights reserved. (NET is the New English Translation.)

Scripture quotations marked (NIV) are taken from the Holy Bible, New International Version®, NIV®. Copyright © 1973, 1978, 1984, 2011 by Biblica, Inc.® Used by permission of Zondervan. All rights reserved worldwide. www.zondervan.com The "NIV" and "New International Version" are trademarks registered in the United States Patent and Trademark Office by Biblica, Inc®.

Scripture quotations taken from the (NASB)® New American Standard Bible®, Copyright © 1960, 1971, 1977, 1995 by the Lockman Foundation. Used by permission. All rights reserved. www.lockman.org.

Scripture quotations marked (ESV) are from the Holy Bible, English Standard Version, Copyright © 2001 by Crossway Bibles, a publishing ministry of Good News Publishers. Used by permission. All rights reserved.

# Contents

*Acknowledgments* ix
*Introduction* xi

## PART A | CREATION AND EVOLUTION, SCIENCE AND THE BIBLE

1. Unraveling Creation and Evolution — 3
2. Science Studies God's Creation — 9

## PART B | THE GOD OF CREATION

3. Creator God Is the Trinity — 15
4. Mind of God before Creation — 23
5. Angels Created before the Earth — 27
6. Perspectives on Creation — 35
7. God's Purpose for Creation — 41

## PART C | DELVING DEEPER INTO CREATION

8. Delving Deeper: How Many Creations? — 47
9. Delving Deeper: How Long Was a Creation Day? — 55
10. Delving Deeper: How Long Have People Been on Earth? — 61
11. Delving Deeper: Creation Myths from Other Cultures — 65

## PART D | CREATION OF THE EARTH

12. Creation Out of Nothing — 75
13. First, Second, Third, and Fourth Days — 81
14. Fifth and Sixth Days: Life in Seas and on Earth — 89
15. Fifth and Sixth Days: Dinosaurs — 95
16. Dinosaur Extinction and Fossils — 105
17. Sixth Day: Ancient Human Ancestors? — 113

18. Sixth Day: Neanderthals … 119
19. Sixth Day: People (Nature and Image of God) … 127
20. God's Nature … 135
21. Manifestations of God … 141
22. God's Image Restored after the Fall … 149
23. Seventh Day: Sabbath … 153

## PART E | OTHER CREATIONS

24. What Else Was Designed and Created? … 161

## PART F | THE THEORY OF EVOLUTION

25. Evolution of Species Basics … 171

## PART G | DELVING DEEPER INTO THEORIES OF EVOLUTION

26. Delving Deeper: More Theories of Evolution (I) … 181
27. Delving Deeper: More Theories of Evolution (II) … 185

## PART H | DELVING DEEPER INTO SCIENCE

28. Delving Deeper: Archaeology Validates the Bible … 193
29. Delving Deeper: Sciences Validate Creation … 199
30. Delving Deeper: Intricate Design Validates Creation … 203
31. Delving Deeper: Genetics Validates Creation … 209
32. Delving Deeper: Genetics and Creation Kinds … 219

## PART I | CREATION OF THE UNIVERSE

33. Creation of the Universe … 225

## PART J | DELVING DEEPER INTO THE UNIVERSE

34. Delving Deeper: Did God Create Aliens? … 235
35. Delving Deeper: The Big Bang Theory … 243

## PART K | CREATION OF NEW HEAVENS (UNIVERSE) AND EARTH

36. Creation of New Heavens (Universe)     251
37. Creation of a New Earth     259
38. Creation of a New Earth: People     265

## PART L | CREATION IS TRUTH

39. What Is Truth?     277
40. Conclusion: It's Up to You!     281

*Prayer to Know God*     287
*Glossary*     291
*Bibliography*     301
*Index*     307

# Acknowledgments

I can't express in words my appreciation for those who worked alongside me during the writing of this book. They have spent many hours reviewing, editing, and providing invaluable feedback to help make this book what it is. Their personal relationship with God and knowledge of the Bible were essential to this end. My wife, Linda, was stalwart in her assistance as she reviewed every chapter. I am also indebted to my Men's Group and the thirteen Beta Readers who spent many hours reviewing and providing input that improved this book.

# Introduction

### YOU CAN FIND THE CREATOR GOD IN THIS BOOK

This book demonstrates that the Creator God (Trinity) of the Bible exists, cares about you, loves you, is available to you, and wants an eternal relationship with you. Look for each person of God as I recount the creation story throughout this book. You can see them everywhere if you choose. It is exciting to know that the supernatural Triune God who created the universe loves you!

### MAJOR THEMES

The major themes about our Creator God (Trinity), creation, and the theory of evolution include:

- There is one Creator God in three persons: the Father, the Son, Jesus Christ, and the Holy Spirit.
- The Intelligent Designer, the Father, used Jesus Christ to speak everything into existence, and the Holy Spirit breathed life into people.
- Jesus continues to sustain all things according to the Father's will.
- There is intelligent design for everything created.
- There are truths about the mysteries, facts, theories, and myths surrounding creation and the theory of evolution.
- Scientific and biblical facts support creation and dispute evolution.
- Genetics, including intelligent cellular design, validate creation and convincingly disprove evolution.
- Nothing has come into existence on its own, no matter the length of time.
- Evolution has always been and will always be an unproven theory.

## CREATED FOR LEARNING

You were created by God for learning. Therefore, this book is created to help you learn what matters to him. This book can be read or studied by individuals or groups. As you will see in this section, it was constructed to benefit a diverse audience through various learning methods. The chapters of this book include narratives, Bible verses, and questions. Some include mysteries, facts, theories, and myths about creation and evolution. These all help separate truth from fiction and enable you to know the Creator God of the Bible.

### Questions to Help You Learn

There are three types of questions included at the end of chapters. "Chapter Questions" are intended to stimulate thinking about specific topics in the chapter. "Personal Application" questions are included to help you think about how a topic relates to you personally. "Dig Deep" questions are a challenge to develop a deeper understanding of a topic. There is extra space for answers to the end-of-chapter questions. This allows you to include your answers when you are thinking about them.

### Special Attention

Pay particular attention when you see the following words: Note, Important, Information, Reminder, and Definition. They provide explanations, definitions, and other information to assist in understanding the text. An Instructional Comment provides further education on the topic at hand. Worldview statements expand a person's view of life and the world in which they live.

### Mystery

This book includes mysteries of God's creations, the theory of evolution, and more. I use the word "mystery" to identify and state these as a question. Mysteries, by their nature, do not have answers.

### Delving Deeper

These 13 chapters include topics that may be difficult to comprehend due to their complex nature. This is true of some information about creation, evolution, and science.

Important: Know what you believe and why you believe it. This book is a learning resource. That's why it includes 279 scripture quotations, 250 questions, many special attention notes and mysteries, 13 Delving Deeper chapters, and 100 researched books and website articles in the bibliography.

**Discovery**

I hope you enjoy your journey through the Bible and the sciences as you discover truths that support the biblical creation account and disprove the various theories of evolution. As they say, "Put your thinking cap on" as you study these fascinating subjects. You will be glad you did!

If you want another book to help you discover more about the God of the Bible, I suggest our first book, *Who Is this God? A Handbook for Life with Him*. It can be purchased on all major online stores that sell books (Amazon, Barnes and Noble, and more).

# PART A

Creation and Evolution,
Science and the Bible

# 1.

# Unraveling Creation and Evolution

Dead dinosaurs don't lie. Neither do dead Neanderthals. Did you know that modern science, genetics, dinosaur fossils, and even Neanderthals support the validity of the Bible and the creation account? The natural world in which people exist did not evolve and come into existence by itself. Even non-Christian scientists have begun to recognize there is design in creation. Christians know everything was created by an Intelligent Designer and spoken into existence by his Son, Jesus Christ.

*Unraveling Creation and Evolution Through Science and the Bible, Exploring Mysteries, Facts, Theories, and Myths* is a Christian nonfiction book that covers a broad landscape of the intersection of faith and science by addressing traditional and less-known aspects of creation and evolution. Problematic subjects are addressed, such as the various theories of evolution, adaptation and natural selection, random and non-random gene mutations, intelligently designed cells, dinosaurs, Neanderthals, the Big Bang Theory, the number of creations, the nature of the Creator God, and much more. Readers explore mysteries, facts, theories, and myths and discover why biblical creation is validated by the Bible and the sciences and why evolution is disproved by its own scientists. As you reflect on the Triune God of the Bible in these pages, you see the Creator God, who always leaves a beautiful picture of his Almighty and loving nature.

## CREATOR GOD

The Creator God described in this book is the Triune God of the Bible, the Father God, his Son, Jesus Christ, and the Holy Spirit. The Father God is the Intelligent Designer who decided how he wanted creation to exist (1 Corinthians 8:6; Hebrews 11:10). Jesus Christ is the person of God who carried out the Father's design when he spoke creation into existence (Colossians 1:15–17). He also sustains creation according to the Father's design and intended purposes. The Holy Spirit is the person of God who breathed the breath of life into people.

## CREATION: BIBLE AND SCIENCE

Without the sciences, we have only a partial, sometimes flawed understanding of the natural world created by the God of the Bible. The Bible and science are not mutually exclusive. Instead, they support one another when reviewed from an objective perspective. The proper understanding of science must come from a biblical foundation. In other words, science helps us understand the Bible, and the Bible helps us understand science.

## WHY READ THIS BOOK?

At the heart of this book is information that describes creation and the various theories of the evolution of the universe and biological species. It will become clear that the Bible and the sciences validate creation and disprove the theories of evolution. Developing a better understanding of these subjects may strengthen your faith in the Creator God of the Bible. As a result, you may find your personal relationship with God enhanced. You may also be better equipped to help others understand these truths more significantly.

### Expanded Worldview

Several chapters contain worldview statements intended to expand the readers' view of themselves and the world in which they live. People develop perspectives, ideas, values, and beliefs primarily derived from the culture in which they live. Together, they create a person's view of themselves and their world. Worldview statements in this book are topical based on a key message in a chapter. They can be about science, the Bible, creation, or evolution. The goal is to help people realign their worldview to that of God.

---

Worldview: The Bible does not have theories and God never guesses because he is omnipotent, omnipresent, and omniscient.

---

### Why Read the Bible?

Multitudes of people have read the Bible. The Bible has been the most-read book in the world for the last fifty years. It has far outsold all other books, with an estimated 4 billion copies sold. I suspect you and everyone you know have at least one version of the Bible. It is a spiritual guide for multitudes of people worldwide. That's because it contains hope, truth, and, most of all, it reveals the one true God of the Bible.

# BIBLE: MYSTERIES, FACTS, THEORIES, AND MYTHS

Scripture contains both facts and mysteries. Most verses in the Bible are statements of fact. However, the Bible also contains mysteries. A mystery in the Bible is a fact whose meaning God has not made clear. Both biblical facts and mysteries are true.

## Biblical Facts

Many verses in the Bible are clear in their meaning. They are easy to understand as biblical facts. We don't need to guess or research to understand them properly.

Two examples of easy-to-understand biblical facts follow:

1. Today you are going to cross the border of Moab, that is, of Ar. (Deuteronomy 2:18 NET)
2. In those days John the Baptist came into the wilderness of Judea proclaiming, "Repent, for the kingdom of heaven is near." (Matthew 3:1–2 NET)

## Biblical Mysteries

There are mysteries in the Bible we can never figure out. The meaning of these mysteries has been concealed from us by God. There are occasions when he does reveal some through divine revelation, as you see below:

> But we speak God's wisdom in a Mystery, the hidden wisdom which God predestined before the ages to our glory. (1 Corinthians 2:7 NASB)

> Now to Him who is able to establish you according to my gospel and the preaching of Jesus Christ, according to the revelation of the Mystery which has been kept secret for long ages past. (Romans 16:25 NASB)

## God Is Greater

God is so much greater than people. We can't fully know what he is saying in a mystery. The mind of God is beyond human understanding.

> "My thoughts are nothing like your thoughts," says the LORD. "And My ways are far beyond anything you could imagine. For just as the heavens are higher than the earth, so My ways are higher than your ways and My thoughts higher than your thoughts." (Isaiah 55:8–9 NLT)

### No Theories in the Bible

A theory is a guess. It may or may not be true. The Bible does not contain theories. As you have just read, it does include facts and mysteries. People create theories about the Bible when they don't understand something in it. So they guess what it means. What's important to know is that the Holy Spirit, the author of the Bible, never guesses about anything. He has no theories.

### No Myths in the Bible

The Bible does not contain myths. Myths are not true. They are unfounded stories created by people who want to explain something they know little about (often about the past). In this book, you will learn about non-biblical myths created by ancient people derived from their own cultures (not the Bible). People still create myths today. These can become popularized, incorporated into culture, and passed on to future generations.

Note: If something is not true, it is false. There is no half-truth.

## CREATOR GOD IS WONDROUS

Within this book, you can find that the Creator God is wondrous and imaginative. Even though each person of God is mysterious, we can know each of them personally. I hope you enjoy your journey of unraveling creation and evolution through the sciences and the Bible.

## CHAPTER QUESTIONS

DQ1: Why can knowledge of the natural world be flawed without some understanding of the sciences?

DQ2. Why do you think so many people have purchased the Bible over the last fifty years?

DQ3. List 3 Bible verses that express a fact.

DQ4: Why does the Bible not contain myths?

## DIG DEEP

DD1: What is the best way to differentiate between a biblical mystery and extrabiblical theories about the Bible?

## PERSONAL APPLICATION

PA1: Why should you pursue understanding mysteries in the Bible, knowing you may not fully understand them?

# 2.

## Science Studies God's Creation

The sciences study the universe and the natural world created by God. Yet, some people think science and the Bible oppose one another. The Bible and science are not mutually exclusive. Instead, they support one another when viewed from an objective perspective. For Christians, the proper understanding of science must come from a biblical foundation.

> The sciences study the universe and the natural world created by God. The Bible and science are not mutually exclusive.

Important: Scientists strive to explain the unknown. As you will see, sometimes, the best explanations end up being theories. Even though theories are not undeniable facts, they can still be valuable references to reality.

### STUDY THE SCIENCES

Every science has its roots in the universe or the natural world God created. During their studies, scientists develop scientific hypotheses, theories, and facts. The scientific method is widely used to separate facts from hypotheses and theories.

### Scientific Method

The scientific method starts with a hypothesis (educated guess). Scientists seeking truth use it to validate that their hypothesis is true. The scientific method is a four-step process involving observations, a hypothesis, experiments, and conclusions.

### Scientific Hypothesis and Theory

Some people mistakenly use "hypothesis" and "theory" as synonyms in science. But they are very different. The following are simple definitions applied to science.

Definition: A *hypothesis* is an idea or assumption that needs to be tested by the scientific method to determine if it's true. It is an "educated guess" that needs to be tested by detailed observations or experimentation.

A *theory* is developed using the scientific method based on evidence from scientific observations and experimentation. It is much more likely to be true than a hypothesis. Yet, evidence is insufficient for the theory to be considered a scientific fact. While theories are not irrefutable scientific facts, many scientists treat them as such.

The following are two well-known scientific theories that you will read about in this book:

- Big Bang Theory: This theory proposes the universe has evolved from a "singularity" event about 13.8 billion years ago.
- Theory of Evolution by Natural Selection: Charles Darwin's theory that all organisms adapt to their environment over time, change their traits and behaviors, and eventually evolve into new species.

### Scientific Fact

A scientific fact is the valid result of repeated testing by the scientific method. It has repeatedly proven to be true under many different conditions. Therefore, objective observations and experimentations repeatedly verify it as a consistent fact.

## MYSTERIES OF SCIENCE

If you look up mysteries of the natural world on the Internet, you will find many that are astounding. There are no scientific explanations for these, but they are scientific facts proven to exist in the natural world.

Following are two incredible examples of mysteries that baffle science:

- Sand dunes: Did you know that thirty-five desert sand dunes across the world sing? They emit a sound like the humming of bees or a rumbling chant. Scientists don't know why this occurs.
- Hessdalen lights: Strange balls of glowing lights over a valley in Central Norway have been seen since at least the 1940s. They come in a variety of colors and formations. Sometimes, they flash, hover, or dart around in the sky. The source of the energy is unknown.

## MYTHS OF SCIENCE

People create scientific myths about something they believe to be true. Myths may occur when people confuse personal experiences with scientific facts. For example, there is a myth that the North Star is the brightest star that can be observed in the heavens. Fact: The brightest visible star is Sirius. Another scientific myth is that lightning never strikes the same place twice. Fact: It can actually strike the same place twice more often than one place once. The Empire State Building is hit by lightning about a hundred times a year.

## YOU CAN KNOW TRUTH

Many people seek answers to questions about life, such as *Who am I? Why am I here? Is there a God? If there is a God, what does he want from me?* We must be careful about who we listen to and what we accept as truth. God warns us to be cautious of those who claim to have knowledge but offer only a human philosophy that is "empty deception."

> See to it that no one takes you captive through philosophy and empty deception, according to the tradition of men, according to the elementary principles of the world, rather than according to Christ. (Colossians 2:8 NASB)

Jesus said we will know the truth when we follow his teachings in the Bible. Knowing truth sets us free from self-delusion and false teachings.

> Then Jesus said to those Judeans who had believed him, "If you continue to follow my teaching, you are really my disciples and you will know the truth, and the truth will set you free." (John 8:31–32 NET)

### Be Vigilant

We must be vigilant to recognize what is true and false. You may find that the science and Bible verses in this book help you to do so. Science, the Bible, and the Creator God of the Bible are in agreement when viewed through an objective lens by those who seek truth.

## CHAPTER QUESTIONS

DQ1: What are the possible results of the scientific method?

DQ2: How can the sciences help Christians understand the Bible better?

DQ3: How can the Bible enable them to understand the sciences better?

DQ4: What is another example of a scientific myth?

## DIG DEEP

DD1: Why does knowing the Bible and some science help Christians differentiate between biblical and scientific facts?

## PERSONAL APPLICATION

PA1: When have you believed a myth about science that led you astray from the truth?

# PART B
The God of Creation

# 3.

## Creator God Is the Trinity

### IN THE BEGINNING, THERE WAS GOD

With these first four words of the Bible, *In the beginning God* (Genesis 1:1), we discover the first recorded biblical doctrine: *there is a self-existent God.* (He's the Triune God of the Bible.) These words bring to life an understanding that everything that exists does so only because the God of the Bible (Trinity) exists and created it. The Bible does not attempt to prove the existence of the Creator God. It simply proclaims from its beginning (Genesis) to its end (Revelation) that the God of the Bible exists.

### NO OTHER GOD EXISTS

In the following verses, God says he alone is God: No other god in any form exists.

> For there is one God. (1 Timothy 2:5 NASB)

> For I am God, and there is no other; I am God, and there is no one like Me. (Isaiah 46:9 NASB)

### ONE PERSON OF GOD OR THREE?

When you see the word "God" in Bible verses as in the above in the Old Testament, it nearly always refers to the Father God, not the Triune God of the Bible (Father God, Son, Jesus Christ, and the Holy Spirit). The Jewish religion of Judaism is a monotheistic religion. They believe there is only one person of God, who is referred to in the Old Testament by many names and titles, such as God, Elohim, YHWH, LORD, Jehovah, and Father.

### Trinity Is in the Old Testament

Even though Jesus and the Holy Spirit are included throughout the Old Testament, the Jewish people do not believe in a Triune God of three persons. As New Testament followers of Jesus Christ, we know that each person of the Trinity was involved in creation. So don't be confused when you read Bible verses and commentaries that refer to "God" as if the Father God was the only one involved in creation (and the Old Testament).

In Genesis 1:1, we see that *Elohim* is the first name given to God in the Bible: It is used in the Hebrew language of the Old Testament 2,570 times.

> In the beginning God *(Elohim)* created the heavens and the earth. (Genesis 1:1 NASB, author's emphasis)
>
> Hebrew *elohim* is the plural form of *El*.

Bible commentators see this as implying the Trinity's involvement in creation (not just the Father). *Elohim* is the infinite, all-powerful God who shows by his works that he is the Creator, Sustainer, and supreme judge of the world.

## GOD OF THE BIBLE IS THE TRINITY

The unfathomable greatness of God can be seen in the triune nature of his existence. The following verse states that the God of the Bible is one God:

> Jesus answered, "The foremost is, 'Hear, o Israel! The LORD our God is one LORD'" (Mark 12:29 NASB)

This one God exists as three persons. This nature of God is referred to as the "Triune God" or the "Trinity." These terms don't appear in the Bible. Biblical scholars created them to help people understand the unity of one God existing in three persons. They are each equally and fully God in every aspect of their essential being and nature.

We see the Trinity mentioned in the following New Testament verses:

> Jesus told his disciples, "Therefore, go and make disciples of all the nations, baptizing them in the name of the Father and the Son and the Holy Spirit." (Matthew 28:19 NLT)
>
> According to the foreknowledge of God the Father, by the sanctifying work of the Spirit, to obey Jesus Christ. (1 Peter 1:2 NASB)
>
> May the grace of the Lord Jesus Christ, and the love of God, and the fellowship of the Holy Spirit be with you all. (2 Corinthians 13:14 NIV)
>
> Mystery: Why is it impossible to fully understand the supernatural nature of the Trinity other than to say it is the three persons of God existing as one God?

# CREATOR GOD IS THE TRINITY

In this book, you will see that each person of the Trinity participated in creation. Therefore, the Trinity is the Creator God of the Bible.

## What Did Each Person Of God Do in Creation?

In brief, the Father God is the Intelligent Designer who decided how he wanted creation to exist (1 Corinthians 8:6; Hebrews 11:10). Jesus Christ is the person of God who carried out the Father's design when he spoke creation into existence (Colossians 1:15–17). He also sustains creation according to the Father's design and intended purposes. The Holy Spirit is the person of God who breathed the breath of life into people (Job 33:4).

---

Each person of the Trinity was involved in creation.
They each had a different role.

---

# FATHER GOD IN CREATION

The following verses tell us the Father God is the Intelligent Designer of creation:

> But we know that there is only one God, the Father, who created everything, and we live for Him. And there is only one Lord, Jesus Christ, through whom God made everything and through whom we have been given life. (1 Corinthians 8:6 NLT, author's emphasis)

In the following verse, we see that everything that exists has a builder and that the Father God is its Builder:

> For every house is built by someone, but the builder of all things is God. (Hebrews 3:4 NET)
>
> *Builder* in Greek is *kataskeuazō*, which means to prepare
>
> thoroughly; by implication, to construct, create.

Another way to think of this verse and the next one is that the Father God is the Intelligent Designer who designed and participated in creating everything, including the "city" of Heaven.

> Abraham was confidently looking forward to a city with eternal foundations, a city designed and built by God. (Hebrews 11:10 NLT)

## The Father Is the Designer of All Things

The following verses suggest the Father God designed and Jesus accomplished his will:

Salvation from the Father through Jesus:

> To the only God our Savior through Jesus Christ our Lord, be glory, majesty, power, and authority, before all time, and now, and for all eternity. Amen. (Jude 1:25 NET)

Peace with the Father through Jesus:

> Therefore, since we have been declared righteous by faith, we have peace with God through our Lord Jesus Christ. (Romans 5:1 NET)

Victory from the Father through Jesus:

> But thanks be to God, who gives us the victory through our Lord Jesus Christ! (1 Corinthians 15:57 NET)

Believers predestined for adoption by the Father through Jesus:

> He did this by predestining us to adoption as his legal heirs through Jesus Christ, according to the pleasure of his will—to the praise of the glory of his grace that he has freely bestowed on us in his dearly loved Son. (Ephesians 1:5–6 NET)

Father God has blessed believers with every spiritual blessing through Jesus:

> Blessed is the God and Father of our Lord Jesus Christ, who has blessed us with every spiritual blessing in the heavenly realms in Christ. (Ephesians 1:3 NET)

## JESUS CHRIST IN CREATION

Like his Father God, Jesus Christ is fully God in every aspect of his divine nature.

> In the beginning was the Word *(Jesus)*, and the Word was with God, and the Word was God. (John 1:1 NASB, author's emphasis)
>
> For in Him all the fullness of Deity dwells in bodily form. (Colossians 2:9 NASB)

In the following verses, we see that the Father created everything through his Son, Jesus Christ:

> In the beginning the Word already existed. The Word was with God, and the Word was God. He *(Jesus)* existed in the beginning with God. God *(Father God)* created everything through Him, and nothing was created except through Him. The Word *(Jesus)* gave life to everything that was created. (John 1:1–4 NLT, author's emphasis)

### Jesus Speaks Creation into Existence

Throughout the creation account in Genesis chapter 1, New Testament believers know that Jesus is the person of God speaking everything into existence.

> God *(Jesus)* said, "Let there be light." (Genesis 1:3 NET, author's emphasis)

Instructional Comment: Moses (and the people of Israel he was writing to) would have thought it was the Father God speaking everything into existence.

God *(Father God)* said, "Let there be light." (Genesis 1:3 NET, author's emphasis)

## Jesus Sustains All of Creation

Jesus not only created everything at the beginning of time, he continues to sustain all that the Father wants to continue.

> He is before all things, and in him all things hold together. (Colossians 1:17 NASB)

> But in these last days he has spoken to us by his Son, whom he appointed heir of all things, and through whom also he made the universe. The Son is the radiance of God's glory and the exact representation of his being, sustaining all things by his powerful word. After he had provided purification for sins, he sat down at the right hand of the Majesty in heaven. (Hebrews 1:2–3 NIV)

> *Sustaining*, Greek is *pherō*; to bear or carry (present tense)

Jesus continues to speak, and everything responds to his words and will. Everything on the Earth remains and continues according to his spoken words. The universe and everything in it, such as new stars forming, old stars dying, black holes forming, etc., respond and obey his words as he speaks the Father's will.

Jesus says in the following verse that the Father God gave him authority over everything, including creation:

> Jesus came and told his disciples, "I have been given all authority in heaven and on earth." (Matthew 28:18 NLT)

## Jesus Fine-Tunes Creation

The following are examples of Jesus' fine-tuning as he sustains the Earth so that it does not self-destruct:

- Sun's distance from the Earth: The sun's surface temperature is over 10,000 degrees. (A sparkler reaches temperatures of 1,800 to 3,000 degrees.) The Earth is 93 million miles from the sun. Significant climate changes would occur if the Earth was permanently 1 inch closer to the sun. Slightly closer than that would result in people burning up all over the Earth. Slightly further away from the sun would result in people freezing all over the Earth.
- Earth's axis and tilt: The Earth spins on its axis with a 23.5-degree tilt. This tilted rotation results in climate variations in different regions of the globe throughout the year. Without this tilt, the climate would be the same year-round in each of the areas of Earth.
- No moon: Life on Earth would be significantly changed without the moon. Without it, nights would be much longer, and days would be much shorter.

Website: The following website details the effects on Earth if we had no moon: https://nineplanets.org/questions/what-would-happen-if-there-was-no-moon

## HOLY SPIRIT IN CREATION

Just like the Father God and Jesus Christ, the Holy Spirit is fully God in every aspect of his divine nature. He has no beginning nor end and lives in eternity. He is omnipotent, omnipresent, and omniscient as they are. He lives in Heaven with the Father and Jesus and in the hearts of every born-again follower of Jesus on Earth. He is their Teacher, Helper, Guide, and friend. Like Jesus, he will never leave or abandon believers.

> I will ask the Father, and He will give you another Helper, that He may be with you forever; that is the Spirit of truth, whom the world cannot receive, because it does not see Him or know Him, but you know Him because He abides with you and will be in you. (John 14:16–17 NASB)

In the following Scripture, we see that the Holy Spirit was involved in creation along with the Father God and Jesus Christ:

> The Spirit of God was hovering over the surface of the waters. (Genesis 1:2 NIV)

> You hide Your face, they are dismayed; You take away their spirit, they expire And return to their dust. You send forth Your Spirit, they are created. (Psalm 104:29–30 NASB)

> The Spirit of God has made me, And the breath of the Almighty gives me life. (Job 33:4 NASB)

## TRINITY IN CONSTANT COMMUNICATION

The three persons of the Creator God are in constant communication. Even though they have their own minds and hearts, they work together in all things. How they do this is a mystery to our limited human minds.

In the following verses, we see that our human understanding and knowledge is limited:

> For we know in part and we prophesy in part, but when the perfect comes, the partial will pass away. When I was a child, I spoke like a child, I thought like a child, I reasoned like a child. When I became a man, I gave up childish ways. For now we see in a mirror dimly, but then face to face. Now I know in part; then I shall know fully, even as I have been fully known. (1 Corinthians 13:9–12 ESV)

The limitations of our human minds and hearts will be removed when we are resurrected with the eternal new body and nature of Jesus Christ. We will see Jesus, the Holy Spirit, and the Father God fully as they truly are. We will no longer need to hide our faces from their full glory as Moses did. This wondrous eternal condition awaits believers in Heaven and the new Earth.

## CHAPTER QUESTIONS

DQ1: Why is each person of the Trinity involved in creation?

DQ2: Why does Jesus need to possess the Father's authority to speak everything into existence and sustain it according to the Father's will and plan?

DQ3: Why is there no place where his authority does not exist?

DQ4: How would you describe the Holy Spirit's involvement in creation?

## DIG DEEP

DD1: How would you describe the difference between the Old Testament idea of one God, the Father, and the New Testament concept of the Triune God of three persons?

## PERSONAL APPLICATION

PA1: Why does it matter to your Christian faith to know the Father God is the Intelligent Designer of creation?

PA2: In what ways are you under the authority of Jesus Christ?

# 4.

# Mind of God before Creation

The Bible does not tell us much about what God was doing before creation. However, we know the Trinity of God (Father, Jesus Christ, and the Holy Spirit) existed before they created the heavens (universe), Earth, and everything in it.

> Before the mountains were born Or You gave birth to the earth and the world, Even from everlasting to everlasting, You are God. (Psalm 90:2 NASB)

> In the beginning was the Word *(Jesus)*, and the Word was with God *(Father God)*, and the Word was God.... All things came into being through Him, and apart from Him nothing came into being that has come into being. (John 1:1–3 NASB, author's emphasis)

## GOD'S MIND AND HEART BEFORE CREATION

The reality that the Triune God is eternal means that they have simultaneously lived in the past, present, and future. God's mind and heart are so much greater than ours. How could we know what they were thinking and planning before creation? The Holy Spirit gives us a glimpse in the scriptures.

---

Worldview: The eternal Triune God's heart and mind are so much greater than people can imagine. His plans were before the beginning of time.

---

The following are examples of what God thought about and planned to do before creation:

The Father God's love for his Son Jesus:

> Father, I want these whom You have given Me to be with Me where I am. Then they can see all the glory You gave Me because You loved Me even before the world began! (John 17:24 NLT)

His plan for Jesus Christ to be the ransom for the salvation of all people:

> God chose Him as your ransom long before the world began, but He has now revealed Him to you in these last days. (1 Peter 1:20 NLT)

His plan to give people his grace for salvation:

> He has saved us and called us to a holy life—not because of anything we have done but because of his own purpose and grace. This grace was given us in Christ Jesus before the beginning of time. (2 Timothy 1:9 NIV)

His plan to give eternal life to his children:

> This truth gives them confidence that they have eternal life, which God—who does not lie—promised them before the world began. (Titus 1:2 NLT)

His plan to give his children an eternal inheritance:

> In Him also we have obtained an inheritance, having been predestined according to purpose who works all things after the counsel of His will. (Ephesians 1:10-11 NASB)

His plan to adopt his children:

> God decided in advance to adopt us into His own family by bringing us to himself through Jesus Christ. (Ephesians 1:5 (NLT)

His plan for his children to receive his glory:

> He does this to make the riches of His glory shine even brighter on those to whom He shows mercy, who were prepared in advance for glory. (Romans 9:23 NLT)

His plan for his children to be holy and blameless (based upon the salvation provided by Jesus):

> Just as He chose us in Him before the foundation of the world, that we would be holy and blameless before Him. (Ephesians 1:4 NASB)

His plan for his children to do good works and to live holy lives:

> For we are His workmanship, created in Christ Jesus for good works, which God prepared beforehand so that we would walk in them. (Ephesians 2:10(NASB)

## BOOK OF LIFE BEFORE CREATION

The Book of Life contains the names of everyone who accepts Jesus Christ as Savior and Lord. Some believe it initially contained the names of everyone who ever lived, but their names were erased if they did not accept Jesus Christ. Others believe it initially contained only the names of those who would accept Jesus Christ.

The following verses describe the Book of Life and the people whose names are written in it.

To the general assembly and church of the firstborn who are enrolled in heaven, and to God, the Judge of all, and to the spirits of the righteous made perfect. (Hebrews 12:23 NASB)

And I ask you, my true partner, to help these two women, for they worked hard with me in telling others the Good News. They worked along with Clement and the rest of my co-workers, whose names are written in the Book of Life. (Philippians 4:3 NLT)

And all the people who belong to this world worshiped the beast. They are the ones whose names were not written in the Book of Life before the world was made—the Book that belongs to the Lamb who was slaughtered. (Revelation 13:8 NLT)

Instructional Comment: The phrase "Book of Life" has Jewish origins referring to an original record of names. This implies that a person's name written in it would never be erased.

Why do you think the Book of Life might be a literal book in Heaven and not just a metaphor for eternal life with God?

## CHAPTER QUESTIONS

DQ1: God lives in eternity, where time and space don't exist and where the past, present, and future are the same. Why do the verses in this chapter affirm that he is eternal and lives in eternity?

DQ2: Why does the Father God love his Son, Jesus Christ?

DQ3: Why is it essential to Christianity that the Father God planned to sacrifice his Son, Jesus Christ for our sins?

DQ4: The Father's plan was for his children to be holy. What does "holy" mean from his perspective?

## DIG DEEP

DD1: Why are God's mind and heart unfathomable for people?

## PERSONAL APPLICATION

PA1: Describe your love for the Father God.

PA2: If you are unsure if your name is written in the Book of Life, now is the time to be sure. Please read the chapter, "Prayer to Know God" and ask him into your heart and life.

# 5.

# Angels Created before the Earth

Angels are supernatural, eternal beings created before anything else. They observed creation and celebrated this work of the Creator God.

> Where were you when I laid the foundations of the earth? . . . What supports its foundations, and who laid its cornerstone as the morning stars sang together and all the angels shouted for joy? (Job 38:4–7 NLT)

The *morning stars* in the above verses are the angels.

Mystery: How long before the creation of the universe and Earth were angels created?

Instructional Comment: The answer to this is not in the Bible. We only know they existed before anything else was created.

## ANGELS ARE CREATED BEINGS

We know from the Bible that angels are created supernatural beings. Since they exist in a spiritual form, they were not made from the natural elements of the Earth. (You will learn about natural elements in the chapter "Creation Out of Nothing.") They were made from spiritual elements that we are not told anything about in the Bible. Even though they are spiritual beings, they can manifest themselves in a visible, physical form.

In the following verse, they are *invisible* creations.

> For by him *(Jesus)* all things were created, in heaven and on earth, visible and *invisible*, whether *thrones or dominions or rulers or authorities*—all things were created through him and for him. (Colossians 1:16, ESV, author's emphasis)

### Angels Belong to the Father God and Jesus Christ

You may be wondering which person of the Triune God the angels belong to. Who has authority over them?

The following verses tell us that they belong to both the Father God and Jesus Christ:

> Regarding the angels, He *(Father God)* says, "He sends **His** angels like the winds, **His** servants like flames of fire." But to the Son He says, "Your throne, O God, endures forever and ever. You rule with a scepter of justice." (Hebrews 1:7–8 NLT, author's emphasis)

> For the Son of Man *(Jesus)* will come with **His** angels in the glory of His Father and will judge all people according to their deeds. (Matthew 16:27 NLT, author's emphasis)

### Hierarchy of Angels

Scripture, like Colossians 1:16, implies a hierarchy of angels. The ranking of angels may be similar to the ranks we see in our human armies.

### Angels Are Supernatural Beings

The Old and New Testaments of the Bible mention the supernatural existence of angels many times. From these, we learn they are diverse in their purposes and functions. Many reside and exercise their supernatural abilities in Heaven, while others do so on Earth and outer space (Daniel 10:12–13, 20).

---

*There is a diversity of angels who have different purposes in serving God. All angels love God.*

---

Scripture tells us there are so many angels that people can't count their number:

> No, you have come to Mount Zion, to the city of the living God, the heavenly Jerusalem, and to countless thousands of angels in a joyful gathering. (Hebrews 12:22 NLT)

They are invisible to people unless the Father sends them in a visible form to Earth on a mission for himself:

> Don't forget to show hospitality to strangers, for some who have done this have entertained angels without realizing it! (Hebrews 13:2 NLT)

The Father God sends his mighty angels to Earth to help people:

> Regarding the angels, He says, "He sends His angels like the winds, His servants like flames of fire." (Hebrews 1:7 NLT)

> Praise the LORD, you angels, you mighty ones who carry out His plans, listening for each of His commands. Yes, praise the LORD, you armies of angels who serve Him and do His will! (Psalms 103:20–21 NLT)

## Armies of Heaven

Angels are mighty warriors referred to as the *armies of Heaven*:

> Suddenly, the angel was joined by a vast host of others—the armies of heaven—praising God. (Luke 2:13 NLT)

These mighty angels will come with Jesus at his second coming:

> When the Son of Man comes in his glory and all the angels with him, then he will sit on his glorious throne. (Matthew 25:31 NET)

# TYPES OF ANGELS

Only a few types of angels are identified in the Bible. The following list represents only those who remained faithful to the Father God and Jesus when the devil led one-third of the angels in rebellion against the Almighty God of the Bible. Demons are not included in this list.

- Archangels (Michael: Daniel 12:1)
- Cherubim (Genesis 3:24)
- Seraphim (Isaiah 6)
- Living Creatures (Revelation 4:8)
- Thrones (Ezekiel 1:15–21)
- Messenger Angels (Not identified as belonging to one of the major types)
- Angels (No type identified; many are just called "angels" in the Bible)

## Types of Angels and Duties

Following are descriptions of the duties of each type of angel:

> Archangels (Jude 1:9; 1 Thessalonians 4:16)
>
> Michael may be the leader of other Archangels. (We don't know from the Bible if there is more than one Archangel.) They are more powerful and mighty than other angels, including the devil and his demons. They are leading warrior angels in charge of protecting people.

Cherubim (Genesis 3:24; Ezekiel 10)

They seem to be powerful and mighty angels responsible for guarding things important to the Father God. For example, the Father God sent them to guard the entry into the Garden of Eden after Adam and Eve were cast out of it. Ezekiel describes them as having four faces (lion, ox, human, eagle).

Seraphim (Isaiah 6:1–7)

They are depicted in Isaiah as having six wings who stand in the presence of the Father God on his throne. They continually say, "Holy, Holy, Holy, is the LORD of hosts, The whole earth is full of His glory." (Isaiah 6:3 NASB)

Living Creatures

They are often associated with Cherubim and appear in the visions of Ezekiel and the Revelation to John. They are described as full of eyes in front and behind, with each of the four living creatures having a different face.

## Individual Names of Angels

There are four angels with individual names in the Bible: Michael, Gabriel, Lucifer, and the Angel of Death.

Michael is called an Archangel in the Bible. Archangel(s) appear to have a leadership role in protecting people and doing warfare against the devil and his demons.

Gabriel is identified as a specific messenger angel carrying the Father God's word to people (Daniel 8:16). He is not mentioned in warfare. When he needed this, he called on Michael to fight (Daniel 10:11–13). Some angels are referred to as "messenger angels" according to Hebrews 1:14. Are messenger angels another type of angel or just a role many angles can perform as directed by the Father God? Gabriel may be one of these messenger angels.

Lucifer is a name for satan (devil) before he led his rebellion against the Father God and Jesus in Heaven. His rebellion and fall are described in Isaiah 14:12–15. The Hebrew for Lucifer is *the morning star*. He is not identified in the Bible as a particular type of angel. He is thought to have been the highest-ranking angel in Heaven before his rebellion.

Angel of Death is an angel used by the Father God in the Old Testament to destroy those who oppose him. We see this in Exodus 12:23 when he was sent to kill the firstborn of everything in Egypt. It was only after this that Pharoah allowed the captive Israelites to leave Egypt and go to their homeland. He seems to be the only angel with these duties.

## WAR IN HEAVEN: DEVIL'S REBELLION

The devil was the most perfect angel (Lucifer), yet he chose to rebel against his Creator. He convinced one-third of the innumerable angels in Heaven to rebel. The result was a war in Heaven that climaxed with the devil and his fallen angels (demons) being cast down to Earth.

> And there was war in heaven, Michael and his angels waging war with the dragon. The dragon and his angels waged war, and they were not strong enough, and there was no longer a place found for them in heaven. And the great dragon was thrown down, the serpent of old who is called the devil and Satan, who deceives the whole world; he was thrown down to the earth, and his angels were thrown down with him. (Revelation 12:7–9 NASB)

Mystery: When was the war in Heaven? How long before the heavens and Earth were created did this war and expulsion to Earth occur?

## TYPES OF DEMONS

There are several lists in the Bible of what could be construed as types of demons. Demons are under the authority of the devil. For example, in Ephesians 6:12, we see the following mentioned:

- Rulers
- Powers
- World forces of this darkness
- Spiritual forces of wickedness in the heavenly places

A thorough search of the Bible will provide only limited information about these. For example, in Daniel 10:11–13, we see the "prince of the kingdom of Persia" identified. He may be one of the "ruling" demons mentioned in Ephesians 6:12. There are also other titles for demons in the Old Testament, such as *Abaddon* (Job 26:6). There are different opinions about whether some "false gods" of the Old Testament refer to demons.

## DEVIL IN GARDEN OF EDEN

The war in Heaven must have occurred before we see the devil tempting Adam and Eve in the Garden of Eden.

> Now the serpent *(devil)* was more shrewd than any of the wild animals that the LORD God had made. He said to the woman, "Is it really true that God said, 'You must not eat from any tree of the orchard'?" The woman said to the serpent, "We may eat of the fruit from the trees of the orchard; but concerning the fruit of the tree that is in the middle of the orchard God said, 'You must not eat from it, and you must not touch it, or else you will die.'" The serpent said to the woman, "Surely you will not die, for God knows that when you eat from it your eyes will open and

you will be like God, knowing good and evil." When the woman saw that the tree produced fruit that was good for food, was attractive to the eye, and was desirable for making one wise, she took some of its fruit and ate it. She also gave some of it to her husband who was with her, and he ate it. (Genesis 3:1–6 NET, author's emphasis)

Jesus told the Jewish people that the devil was a murderer and liar by his fallen nature. When he lied in the Garden, he did so out of his twisted hate for the God of the Bible and people.

> For you are the children of your father the devil, and you love to do the evil things he does. He was a murderer from the beginning. He has always hated the truth, because there is no truth in him. When he lies, it is consistent with his character; for he is a liar and the father of lies. (John 8:44 NLT)

> Mystery: How long were Adam and Eve in the Garden of Eden before their rebellion and fall? Before Adam and Eve's fall, how long was the serpent, the devil, in the Garden of Eden?

Instructional Comment: There are many mysteries about creation for which we cannot know the answers. However, as followers of Jesus Christ, we stand firm in our faith and trust in our Father God, Jesus Christ, the Holy Spirit, and their inerrant word in the Bible.

## JESUS IS SUPERIOR TO ANGELS

The Father God declares that his Son, Jesus Christ, is superior to angels (and the devil and demons, which were once angels).

> Having become as much superior to angels as the name he has inherited is more excellent than theirs. For to which of the angels did God ever say, "You are my Son, today I have begotten you"? Or again, "I will be to him a father, and he shall be to me a son"? And again, when he brings the firstborn into the world, he says, "Let all God's angels worship him." (Hebrews 1:4–6 ESV)

In Colossians 2:16–23, we are told not to worship angels. Even though they are mightier than people, they are not Almighty God. Only the Triune God deserves our worship (Exodus 20:3–5).

> Let no one who delights in false humility and the worship of angels pass judgment on you. That person goes on at great lengths about what he has supposedly seen, but he is puffed up with empty notions by his fleshly mind. (Colossians 2:18 NET)

Remember that God's angels are his servants, not the servants of people.

## CHAPTER QUESTIONS

DQ1: How would you describe an angel to someone unfamiliar with them?

DQ2: Why are angels called the "armies of heaven"?

DQ3: Why are demons called "fallen angels?"

DQ4: Which type of angel is the highest ranking?

## DIG DEEP

DD1: Why must angels and demons be organized as armies with various ranks of leaders (like a human army)?

## PERSONAL APPLICATION

PA1: When have you seen an angel, experienced one helping you, or heard about one helping someone else?

# 6.

## Perspectives on Creation

One purpose of this book is to help you understand that creation is the story of our loving and almighty Creator God as the Father, Son, Jesus Christ, and the Holy Spirit. As we seek to understand creation, we will discover who the Bible's Creator God is. Please remember this as you read the various perspectives of people about creation.

### TRADITIONAL VIEW OF CREATION

I refer to this traditional view as "creationism." It's the view that many Christians have about the creation account. Most believe that Genesis chapters 1 and 2 represent one account of the initial creation, while each of these chapters provides unique details. Chapter 1 describes the six-day creation process. Chapter 2 provides more information on the creation of plant life, Adam and Eve, the Garden of Eden, and more. Its perspective is on Adam, his charge to tend the plant life, name the animals, and the need for a help-mate in Eve. It reflects the need for a man and a woman to complete each other and work together in life.

Definition: *Creationists*: Believe in the traditional creation account described in Genesis chapters 1 and 2. I include myself and others here who believe the biblical truths about creation but don't necessarily support all Young or Old Earth Creationists' ideas.

*Young Earth Creationists*: Believe in the creation account in Genesis and believe strongly that a creation day is a literal 24 hours. Based on this, they theorize the Earth, all living creatures, and the universe are about 6,000 years old (perhaps as old as 10,000 years).

*Old Earth Creationists*: Believe in the creation account in Genesis but do not support the idea that a creation day was 24 hours. They believe a creation day might have been a longer period. Therefore, they reject the theory that the Earth, the universe, and everything living are 6,000 to 10,000 years old. Instead, they believe these are much older. However, they are not evolutionists.

The traditional view of creationism assumes the following are true:

- Genesis 1:1: The universe, including Earth, was created out of nothing
- Genesis 1:2: The Earth was formless and empty of life after its creation
- Genesis 1:1–2: Summary of the creation account elaborated in Genesis chapters 1 and 2
- Genesis chapters 1 and 2: The Earth was made habitable, and all life was created

## Facts about Creation

A biblical fact is true. Noting is false in the Bible. The Bible and the account of creation are fact-based, not theories.

The following are what I believe to be biblical facts about creation:

- Father God is the Intelligent Designer of all creation
- Jesus Christ, following the Father's will, spoke the universe, Earth, and all living things into existence
- The universe and Earth were created out of nothing (*Ex Nihilo*)
- After creating the universe out of nothing (Genesis 1:1), three creation days were used to make the Earth habitable, and then three days were used to create everything living by their kind
- Creator God rested on the seventh day from the work of creation
- Genesis chapter 1 describes the creation activities for the six days
- Genesis chapter 2 provides additional specific details about creation, including man and woman, the Garden of Eden, the provision of water to cause the growth of plant life, and more

## Theories about Creation

Someone asked, "How old is the Earth?" The reply was, "Older than dirt!" But what does this explanation have to do with facts and theories about the creation account?

Some people might get nervous when they hear the word "theories" when referring to the Bible and creation. I hope you will see why I refer to some ideas about creation as facts and others as theories. *A theory about creation and the Bible is an explanation suggested by people who want to enhance their understanding of the Bible.* The age of the Earth is one example of such a theory.

> A theory about creation or the Bible is an explanation suggested
> by people who want to enhance their understanding of the Bible.
> Theories are not facts. There are no theories in the Bible.

Following is a list of beliefs I consider unproven theories about creation:

- Young Earth Creationism age of the universe, Earth, and people on the Earth is between 6,000 and 10,000 years (based on a 24-hour creation day). Universe was created to be instantly mature and functional.
- Each of the six creation days was a 24-hour day.
- Old Earth Creationism age of the universe is 13.7 billion years. Jesus created and guided the universe into its current condition.
- Each of the six creation days was an indefinite period.
- The time between each creation day was longer than 24 hours (even if the creation day was 24 hours).

One idea we will explore in the chapter, "First, Second, Third, and Fourth Days," is whether each "creation kind" was a super-genetic species from which many species or varieties developed over time.

## CREATION IS CHRISTIAN DOCTRINE

Creation is a fundamental doctrine (teaching) of the Christian faith. Without it, nothing exists. It's our first glimpse of the loving, Almighty Creator God of the Bible. Genesis chapters 1 and 2 teach how Christians should live and interact with God's created world.

## HAVE WE LOST A FULL BIBLICAL VIEW OF CREATION?

Edward W. Klink III, in his book *The Beginning and End of All Things, A Biblical Theology of Creation and New Creation*, says we have. Many people no longer consider God's creation story foundational to the Bible. Instead, they have relegated it as unimportant, a subtitle on discussions about origins. Klink says that a proper biblical understanding of the origins of the universe, the natural world, all living creatures, and people is paramount to believing in the God of the Bible. God is much more than what we read in Genesis chapters 1 and 2. Yes, the creation story is about God's initial creation, but it is also about his later new creation and everything between them, as discovered throughout the entire Bible.

Edward Klink shows how the creation story has lost much of its meaning in the life of Christians and Christ's church today. He focuses on four reasons Christians have lost their holistic view of God, creation, and the Bible. The result, he says, is denying or distorting key aspects of the full biblical creation story and the impact it should have as we study the entire Bible.

Note: Edward Klink, like many of the authors and commentators in this book, refers to "God" in discussions of creation. By now, you know that the Creator God of the Bible is the Trinity that worked together to create everything.

## CREATION APOLOGETICS

The term "apologetics" generally means the defense of the Christian faith, including the gospel of Jesus Christ. Its purpose is to answer non-believers' and believers' questions about the Christian faith. When applied to the topic of creation, it is referred to as "creation apologetics."

Creation apologetics describes how to corroborate the accuracy of the Bible's account of creation, contradict the ideas of evolution, and help people find biblical answers to questions about God, why they are here, and much more.

### Three Creation Apologetics Websites

The authors and information presented on the following three Christian websites are Young Earth Creationists. Their purposes are to inform people and defend the Christian faith with creation apologetics. They provide many sound biblical explanations along with science about important questions people often have. They support the theory that the universe, Earth, and everything living is about 6,000 years old (some think up to 10,000 years).

> *Answers in Genesis* is a good Young Earth Creationists website for researching biblical beliefs grounded in scientific analyses. Ken Ham has been a leader in creation apologetics for decades. He has had many encounters and debates with those who support evolution. He provides biblical and scientific answers about the book of Genesis and creation in particular.

Website: You can learn more about Answers in Genesis on the following website: https://answersingenesis.org

> *Genesis Apologetics* is a Young Earth Creationists website for people of all ages to learn about creation and why it is more logical than evolution. Their analyses and presentations of biblical truth are examples of creation apologetics. The website offers many short answers to common and unique questions. This includes questions and answers about the flood, various anthropological discoveries declared to be ancient human ancestors, dinosaurs, and much more.

Website: You can learn more about Genesis Apologetics on the following website: https://genesisapologetics.com/creation-vs-evolution-fast-facts

> *Is Genesis History* is another good Young Earth Creationists website to learn about creation and how modern science validates creation and disproves evolution. Dr. Del Tackett has created a sophisticated documentary film entitled "Is Genesis History?" Various scientists provide science-based evidence from archaeology, geology, genetics, paleontology, and other scientific disciplines to support the creation account and their Young Earth Creation theory.

Website: You can learn more about Is Genesis History on the following website: https://isgenesishistory.com/free-videos-is-genesis-history/

Note: Whether or not you subscribe to the Young Earth Creationists' view of humanity's age, you will find compelling evidence to support God's story of creation on their websites.

## BIBLE, ATHEISM, AND SCIENCE

A Time magazine article, "Is God Dead?" in 1966, brought atheism to the forefront of American thought. It became trendy to believe there was no God. However, since then, the idea that God does not exist has become less popular. Eric Metaxas's book, *Is Atheism Dead?* is based on observations that atheism is far less acceptable than before. Today, we see that the sciences validate the creation story and, therefore, the existence of God. Eric believes that many people today see atheism as irrational. So, he says that atheism is dead, not God.

    Eric addresses much more than atheism. He brings together science and the Bible to explain the created universe. He addresses modern archaeological advancements, historical evidence, and testimonies from scientists in his detailed analyses. His thirty topical chapters clearly and logically engage readers in the wonder of God as the Intelligent Designer of everything and the universe in particular. His chapters include the "Big Bang Theory," the origin of life, the failure of evolution to explain our world, and much more. He sees the universe and our planet Earth as being "fine-tuned" by the Creator God. For example, this fine-tuning enables Earth to be inhabitable for all living organisms, particularly people.

## CHAPTER QUESTIONS

DQ1: What do you know about the biblical account of creation?

DQ2: Does it align with what you have read in this chapter? If not, why?

DQ3: Why is creation considered Christian doctrine?

DQ4: How do you define Christian apologetics?

## DIG DEEP

DD1: How would you use creation apologetics to a non-believer? What would you say?

## PERSONAL APPLICATION

PA1: If someone asked if you believed the creation account in the Bible, could you describe it to them accurately?

PA2: Why would this question allow you to respond from a creation apologetics view?

# 7.

# God's Purpose for Creation

The creation account demonstrates the Creator God's majesty, unique power, and all-encompassing authority to create anything and everything he desires. This chapter helps us understand the purposes of creation.

## CREATION POINTS TO THE GLORY OF THE CREATOR GOD

We know that God is the Triune Creator God of the Bible. But what is God's glory? God's glory must be an aspect of his divine nature. His glory is a radiant display and reflection of his majesty and holiness. Perhaps glorifying God is praising him for who he is and what he does.

Instructional Comment: No one can see the full radiance of God's glory and live (Exodus 33:18–23).

The following verses point to the glory of the Creator God as we view his creative work:

> The heavens declare the glory of God, and the sky above proclaims his handiwork. (Psalm 19:1 ESV)
>
> Instead, glorify His mighty works, singing songs of praise. (Job 36:24 NLT)

## CREATION DEMONSTRATES THE CREATOR GOD'S LOVE

Everything the Triune God of the Bible does emerges from a heart of divine love.

> We have come to know and have believed the love which God has for us. *God is love* and the one who abides in love abides in God, and God abides in him. (1 John 4:16 NASB, author's emphasis)

Christians often consider the Greek word used here for love, *agape̅* as referring to the Creator God's unmerited, unconditional love for people. This love is freely given and reflects his compassion, mercy, and grace. No one or nothing deserves this unfailing love that was demonstrated in creation.

Can you imagine that the universe (planets, solar systems, and other cosmic objects) was created out of love? But why would the Creator God love these lifeless objects in space where no people exist (except on Earth)? Perhaps the answer is that God loves to create.

### How Different from Loveless, Heartless Evolution

Evolution is about the physical world and survival of the fittest. It is a theory that cannot explain the soul and personal relationships. There is no place for love and compassion in the theory of evolution. It's all about undirected chance without the guiding hand of a loving Creator God.

---

> Worldview: Evolution is about physical changes. It is a theory that cannot explain the soul and is not a proven scientific fact.

---

## EARTH: DWELLING PLACE FOR GOD'S INCARNATE SON

The Father designed, and Jesus made the Earth a suitable physical environment for him to come as the Son of Man in a physical human form.

> So the Word became human and made His home among us. He was full of unfailing love and faithfulness. And we have seen His glory, the glory of the Father's one and only Son. (John 1:14 NLT)

## EARTH: DWELLING PLACE FOR GOD'S PEOPLE

The Father designed, and Jesus made the Earth a suitable physical environment for people to live, populate, and enjoy in a physical form. The Earth was created for people, not the supernatural God, to dwell in.

> For the LORD is God, and He created the heavens and earth and put everything in place. He made the world to be lived in, not to be a place of empty chaos. (Isaiah 45:18 NLT)

> So God created human beings in His own image, . . . Then God blessed them and said, "Be fruitful and multiply. Fill the earth and govern it." (Genesis 1:27–28 NLT)

## EARTH: SUITABLE PLACE FOR PEOPLE TO LIVE AND WORK

Since the Earth was created to be a suitable place for people to live, it would undoubtedly be made an appropriate place for people to work.

> The LORD God took the man and put him in the garden of Eden to work it and keep it. (Genesis 2:15 ESV)

> When I look at the night sky and see the work of Your fingers—the moon and the stars You set in place—what are mere mortals that You should think about them, human beings that You should care for them? . . . You gave them charge of everything You made, putting all things under their authority. (Psalm 8:3–6 NLT)

## CHAPTER QUESTIONS

DQ1: How would you describe the glory of God to people who do not know him?

DQ2: Which of God's purposes for creation do you think is the most important? And why?

DQ3: Why did the Father design the Earth to be suitable for his incarnate Son to come?

DQ4: What would happen if the Earth's physical environments were not created as suitable places for people to live?

## DIG DEEP

DD1: Why does God have a purpose in everything he does, including creation?

## PERSONAL APPLICATION

PA1: Why is the creation of planet Earth a suitable place for you personally to live and work?

PA2: How does seeing the universe encourage you to praise the Creator God of the Bible?

# PART C
Delving Deeper into Creation

# 8.

# Delving Deeper: How Many Creations?

Important: Delving Deeper chapters contain material that may be difficult to understand. I suggest you spend extra time studying it and ask for the Holy Spirit's help. The Father and Jesus sent him to live within Christians as their Teacher, Guide, and Helper.

## TWO CREATION ACCOUNTS: INITIAL AND NEW CREATION

The Bible clearly describes two creation accounts: the initial creation in Genesis chapters 1 and 2 and the new creation of the heavens and Earth in Revelation chapters 21 and 22.

### Traditional One Initial Creation Account

In the traditional view, Genesis chapters 1 and 2 represent one account of the initial creation, with each chapter providing unique details. Chapter 1 describes the six-day creation process. Chapter 2 provides additional information on the creation of plant life, Adam and Eve, the Garden of Eden, and more. Its overall perspective is on Adam, his charge to tend the plant life, name the animals, and the need for a help-mate in Eve. It reflects the need for a man and a woman to complete each other and work together. Also, chapter 2 is not as sequential as is chapter 1.

### Rebellion and Corruption

When Adam and Eve rebelled in the Garden of Eden, their sin resulted in the corruption of all creation, including their nature (therefore, the nature of all people).

> Against its will, all creation was subjected to God's curse. But with eager hope, the creation looks forward to the day when it will join God's children in glorious

freedom from death and decay. For we know that all creation has been groaning as in the pains of childbirth right up to the present time. (Romans 8:20–22 NLT)

The corruption of human nature resulted in the corruption of the image of God within people. This corruption led to spiritual death, which means that all people are born separated from the Triune God with an eternal destiny of the Lake of Fire (John 3:18). Only salvation in Christ transfers people from this fate to eternal life with the God of the Bible.

> When Adam and Eve were created in God's image, that image was perfect, without defects. Their rebellion against God lead to the corruption of human nature and all creation.

### New Heavens (Universe) and Earth Creation Account

The corruption of everything initially created is why a new creation is required at the end of time. The Almighty God of the Bible destroys the corrupted heavens (universe) and Earth by fire (2 Peter 3:7–10). A new heavens (universe) and Earth are created as the eternal, incorruptible home for all believers in Christ.

> Then I saw a new heaven and a new earth, for the first heaven and the first earth had passed away, and the sea was no more. And I saw the holy city, new Jerusalem, coming down out of heaven from God, prepared as a bride adorned for her husband. And I heard a loud voice from the throne saying, "Behold, the dwelling place of God is with man. He will dwell with them, and they will be his people, and God himself will be with them as their God. He will wipe away every tear from their eyes, and death shall be no more, neither shall there be mourning, nor crying, nor pain anymore, for the former things have passed away." And he who was seated on the throne said, "Behold, I am making all things new." Also he said, "Write this down, for these words are trustworthy and true." (Revelation 21:1–5 ESV)

Born-again Christians look forward to this new creation as an eternal blessing of life in the presence of the Father God, Jesus Christ, and the Holy Spirit. The corruption is destroyed, so there will no longer be any pain, suffering, sorrow, death, or sin.

Note: Some people have suggested the traditional account of just one initial creation may be incorrect. Some believe in one of the following theories about two initial creations.

## TWO INITIAL CREATIONS: GENESIS 1:1 GAP THEORY

People who support this theory believe there were two distinct initial creations described in Genesis chapter 1. First Initial Creation: They think the universe was made over a lengthy period (Genesis 1:1–2), which left the Earth formless and empty. Second Initial Creation: They believe the second creation starts with Genesis 1:3 and includes the remaining verses of Genesis chapters 1 and 2. The gap is between the end of the first creation (Genesis 1:1–2) and the start of the second creation (Genesis 1:3). This theory has several variations.

**Earth Was Formless and Void Theories**

Those who believe in the Gap Theory think there are two theories as to why the Earth was formless and void of life after Jesus created the universe and Earth in Genesis 1:1.

Following are two theories about why the Earth was formless and empty of life:

- One: The Earth was intentionally left unfinished after the universe's creation
- Two: Something occurred that caused the Earth to be devastated after it and the universe were created

Bible commentator Adam Clarke[1] says the following in supporting the first theory that Earth was intentionally left unfinished after the universe's creation:

> The earth was without form and void—The original terms והת tohu and ובה bohu, which we translate without form and void—convey the idea of confusion and disorder (chaos). God seems at first to have created the elementary principles of all things; and this formed the grand mass of matter, which in this state must be without arrangement, or any distinction of parts: a vast collection of indescribably confused materials, of nameless entities strangely mixed. When this congeries of elementary principles was brought together, God was pleased to spend six days in assimilating, assorting, and arranging the materials, out of which he built up, not only the earth, but the whole of the solar system. (Adam Clarke's Commentary of the Bible)

Information: Adam Clarke (1762 to 1823) was a British theologian, pastor, and Bible scholar, commentator. He is one of several commentators I use for Bible studies.

> Mystery: Why was the Earth formless and empty of life, and how long did it exist in that state?

## TWO INITIAL CREATIONS: GENESIS CHAPTERS ONE AND TWO CONTRADICTION THEORY

This theory is based on the supposition that Genesis chapters 1 and 2 contradict one another in their accounts of when plant life and Adam and Eve were created.

Instructional Comment: The God of the Bible doesn't use contradictions in the Bible. Any suggested contradiction is the result of our inability to fully understand the meaning of the Bible.

---

1. Clarke, *Commentary*, Genesis 1:1.

### Contradiction of Plant Life?

The issue surrounding this theorized contradiction is when plant life came into existence: Was it in chapter 1 when it was created, or in chapter 2 when the ground was watered, and it sprouted when there was someone to tend the land?

Chapter 1: Genesis 1:11–13 says plant life was created on the third day:

> Then God *(Jesus)* said, "Let the land produce vegetation: seed-bearing plants and trees on the land that bear fruit with seed in it, according to their various kinds." And it was so. The land produced vegetation: plants bearing seed according to their kinds and trees bearing fruit with seed in it according to their kinds. And God saw that it was good. And there was evening, and there was morning—the third day. (Genesis 1:11–13 NIV, author's emphasis)

Chapter 2: Genesis 2:5–9 says that plant life did not grow because the ground was not yet watered and there was no one to take care of the land and its plant life:

> Now no shrub had yet appeared on the earth and no plant had yet sprung up, for the LORD God had not sent rain on the earth and there was no one to work the ground, but streams came up from the earth and watered the whole surface of the ground. Then the LORD God formed a man from the dust of the ground and breathed into his nostrils the breath of life, and the man became a living being. Now the LORD God had planted a garden in the east, in Eden; and there he put the man he had formed. The LORD God made all kinds of trees grow out of the ground—trees that were pleasing to the eye and good for food. (Genesis 2:5–9 NIV, author's emphasis)

Instructional Comment: Perhaps in chapter 1, Jesus created the seeds of all plant life and placed them in the ground. Then, in chapter 2, the seeds germinated and sprouted once they received water, and Adam was there to attend to their growth.

### Contradiction of Adam and Eve?

The issue surrounding this theorized contradiction is when and how Adam and Eve were created.

Chapter 1: Genesis 1:26–27 says that the first "human beings," Adam and Eve, were created in God's image on Day Six. Male and female, they were created:

> Then God said, "Let Us make human beings in Our image, to be like Us. They will reign over the fish in the sea, the birds in the sky, the livestock, all the wild animals on the earth, and the small animals that scurry along the ground." So God created human beings in His own image. In the image of God He created them; male and female He created them. (Genesis 1:26–27 NLT)

Instructional Comment: In the above verses, we see Jesus speaking for the Trinity, saying they will make people in their image.

Chapter 2: Genesis 2:7, 2:18, and 2:21–23 provide more definitions about their creation:

> Then the LORD God formed a man from the dust of the ground and breathed into his nostrils the breath of life, and the man became a living being. (Genesis 2:7 NIV)

> Then the LORD God said, "It is not good for the man to be alone. I will make a helper who is just right for him." (Genesis 3:18 NLT)

> So the LORD God caused the man to fall into a deep sleep; and while he was sleeping, he took one of the man's ribs and then closed up the place with flesh. Then the LORD God made a woman from the rib he had taken out of the man, and he brought her to the man. The man said, "This is now bone of my bones and flesh of my flesh; she shall be called 'woman,' for she was taken out of man." (Genesis 2:21–23 NIV)

Instructional Comment: Chapter 1 says Adam and Eve were created in God's image. Chapter 2 focuses on how Jesus created each of them. It also adds more information about why he created the woman. She was designed to be the husband's friend, co-equal, co-ruler over the Earth, and marriage partner to bear children to populate the Earth (Genesis 1:27–28).

## OTHER PROMINENT CREATION THEORIES

The following two theories come from the Young Earth Creationist and Old Earth Creationist interpretations of the initial creation account in Genesis chapter 1.

### Young Earth Creationism

Young Earth Creationists believe in a literal interpretation of the creation account in Genesis chapter 1, including a 24-hour day for the six days of creation. Based on this, they theorize that the Earth and all living creatures, including people, and the universe, are about 6,000 years old (perhaps up to 10,000 years). Because of the 24-hour creation day, they think all life was created instantly, fully mature, and ready to propagate new life. They believe Adam and Eve were distinctly made in God's image. They believe the universe was instantly made (out of nothing) to be fully mature and functional. They see the geologic record, including the flood, as supporting their age of the Earth view. Ken Ham (Ark Encounter) and Del Tackett (Is Genesis History?) promote this theory. They provide considerable biblical, historical, and scientific evidence to support their beliefs.

Website: More can be learned about Young Earth Creationism from the following website: https://www.christianity.com/wiki/bible/what-is-young-earth-creationism.html

### Old Earth Creationism

Old Earth Creationists believe in a literal interpretation of the creation account in Genesis chapter 1. Even though their beliefs in creation are similar to Young Earth Creationists, they differ primarily in terms of time. They believe the Earth, all living creations, and the universe could be millions or billions of years old, not 6,000 years. This is based on the possibility that the six days of creation were not literal 24-hour days but rather longer periods.

They think Jesus created the universe miraculously out of nothing but may have allowed it to mature over long periods. They believe Adam and Eve were distinctly made in God's image. The age of the universe is thought by many Old Earth Creationists to be 13.7 billion years old and Earth to be millions or billions of years old. However, they are not evolutionists and do not support Darwin's theory of biological evolution of the species. "Day-Age Creationism" and "Progressive Creationism" are controversial variations of this theory.

Website: More information about Old Earth Creationism can be found on the following website: https://www.scienceandfaith.org/old-earth-creationism

## CHAPTER QUESTIONS

DQ1: Why do you think the new heavens and Earth will be eternally perfect and incorruptible?

DQ2: Why do you think these verses in chapters 1 and 2 are not contradictions that result in two initial creation accounts about plant life and Adam and Eve?

DQ3: Why do you think believing in the traditional view of creation in Genesis chapters 1 and 2 is sufficient and that the issue of creation time is not as important as what Jesus did on each creation day?

DQ4: Describe how Young Earth Creationism and Old Earth Creationism are the same and different.

## DIG DEEP

DD1: Describe why geology is interpreted differently by Young Earth Creationists and Old Earth Creationists advocates.

## PERSONAL APPLICATION

PA1: Are you a Young Earth Creationist or an Old Earth Creationist? And why?

# 9.

# Delving Deeper: How Long Was a Creation Day?

Important: Delving Deeper chapters contain material that may be difficult for some to understand. I suggest you spend extra time studying it and ask the Holy Spirit to help you. The Father and Jesus sent him to live within Christians as their Teacher, Guide, and Helper.

This chapter examines the use of the Hebrew word for day in the Old Testament and the creation account in Genesis. My intent is to provide a biblically accurate understanding of the word without championing one perspective. It's up to you to decide what you believe.

## YOUNG EARTH AND OLD EARTH CREATIONISTS' DEBATE

There has been debate for decades among creationists, especially between Young Earth and Old Earth Creationists, about the duration of a creation day. The issue is whether each day is literally twenty-four hours or an extended period. At the center of this debate is the meaning and appropriate use of the Hebrew word (*yôm*) for day used in Genesis and the Old Testament.

---

*We will never know which creationist view is correct about the length of a creation day. What is essential is understanding what Jesus created each day.*

---

## HEBREW DEFINITION OF DAY

Those who believe in a 24-hour creation day view the book of Genesis as history. If it is history, then a 24-hour day may be correct. However, other alternative views propose a longer creation day. Some say the six days of creation in Genesis chapter 1 can be studied and understood from a descriptive view. In other words, the intention of these verses may have been to help people know what God did, not literally how long it took.

Important: The book of Genesis must be studied as a historical narrative because biblical history is factual and literal. If not, we would falsely question whether Abraham and King David existed. Did the flood really occur, or was it a myth? Biblical text should always be interpreted as literal. Only if context determines otherwise should it be understood as figurative (such as figures of speech, poetry, parables, and so on). For example, it is a historical fact that Jesus Christ came to Earth. Within the factual context of the four Gospels (Matthew, Mark, Luke, and John), Jesus used many parables to explain spiritual concepts. The parable is not a literal historical fact, even though it was told in a historical setting. Some see Hebrew figures of speech in the literal account of the six days of creation, which indicates that time is not literal but figurative.

### Definition of a Hebrew Day

The Hebrew word for day, *yôm*, used in the creation account, carries multiple meanings in the book of Genesis and the Old Testament.

*Day*, Hebrew is *yôm*; It's translated with literal and figurative meanings.

## USES OF HEBREW WORD FOR DAY

A study of the Old Testament Hebrew text reveals this versatile Hebrew word *yôm* is used 2,301 times in the Old Testament with various meanings. All of these uses are time-related.

Note: The significant differences in the uses of *yôm* cause us to pause and consider how it impacts the meaning of a verse. It is not always used to represent a literal 24-hour day.

### Hebrew Day in Creation Account

The Hebrew word *yôm* is used frequently in the creation account in the Bible. Let's look at these to see how they might impact our understanding of a creation day.

12-Hour Time Period:
Evening and morning in each of the six days of creation can refer to the 12-hour nighttime period. Daytime would be another 12-hour period.

> God called the light day, and the darkness He called night. And there was evening and there was morning, one day. (Genesis 1:5 NASB)

Instructional Comment: In the Jewish calendar, a new day starts at 6 pm in the evening. The morning starts at 6 am.

24-Hour Day:
Evening and morning refer to a total period of 24 hours.

> Then God *(Jesus)* said, "Let there be lights in the expanse of the heavens to separate the day from the night, and let them be for signs and for seasons and for days and years." (Genesis 1:14 NASB, author's emphasis)

Entire Creative Week:
The seven days of creation are summarized as a day.

> This is the account of the heavens and the earth when they were created, in the day that the LORD God made earth and heaven. (Genesis 2:4 NASB)

Finite Cycle of Time:
Evening and morning together with an ordinal/cardinal number (one day) can represent a day with any finite period. Therefore, each day could be a separate cycle of an unknown length. So, each day of creation is a separate cycle in the total cycle of six days.

> God called the light day, and the darkness He called night. And *there was evening and there was morning*, one day. (Genesis 1:5 NASB, author's emphasis)

Note: The preceding information and study of the Hebrew word for day are abstracted from the article "Word Study—Yom" by Greg Neyman, first published on March 16, 2005. I chose this Old Earth Creationist website because it offered considerable details of the biblical usage of the word *yôm*: https://www.oldearth.org/word_study_yom.htm.

## EVENING AND MORNING

Young Earth Creationists say that every occurrence of evening and morning together in the Bible (38 times of the 2,301 occurrences of *yôm*) always refers to a twenty-four-hour day. However, this phrase is used in Daniel 8:26 to indicate a long period (not 24 hours).

> The vision of the evenings and mornings that was told to you is correct. But you should seal up the vision, for it refers to a time many days from now." (Daniel 8:26 NET)

### Merismus Hebrew Figure of Speech

The word *merismus* in biblical Hebrew represents a Hebrew figure of speech that uses two contrasting parts to express totality or completeness. The focus of the pairs is not the individual parts but rather the idea of the whole. *We see this in the Genesis days of creation with the merismus expression, "there was evening and there was morning." This phrase refers to the whole period regardless of how long each part or the whole is.*

### Other Examples of the Use of Merismus

The "Old Testament" and "New Testament" are a *merismus* representing the totality of the entire Bible regardless of how long are the two parts or the entire Bible.

There are additional examples of *merismus* pairs in the Old Testament. In the following verses, Moses uses a figure of speech when he tells people to continually teach, talk about, and keep Scripture in their minds. The *merismus*, or the whole idea, is that they must constantly speak and think about Scripture regardless of how much time is spent doing this.

> These words, which I am commanding you today, shall be on your heart. You shall teach them diligently to your sons and *shall talk of them when you sit in your house and when you walk by the way and when you lie down and when you rise up. You shall bind them as a sign on your hand and they shall be as frontals on your forehead.* (Deuteronomy 6:6–8 NASB, author's emphasis)

In the next verses, the Psalmist says that no matter where he is or what he is doing, the God of the Bible always knows everything about him. The *merismus* pairs are *I sit down* and *I rise up*, and *my path* and *my lying down*.

> O LORD, You have searched me and known me. You know when *I sit down* and when *I rise up*; You understand my thought from afar. You scrutinize *my path* and *my lying down*, And are intimately acquainted with all my ways. (Psalm 139:1–3 NASB, author's emphasis)

## English Uses of Merismus

We use this same figure of speech pattern in English when we say such things as "*in sickness and health*" and "*come rain or shine*" to represent the entirety of marriage or all weather conditions regardless of how much time is involved. For example, the phrase "*in sickness and health*" means a man and woman are committing to remaining with each other for the entire marriage, no matter what happens to them. The phrase "*ladies and gentlemen*" is a *merismus* that refers to all the people without regard to the number of people.

## Other Considerations about the Length of a Creation Day

Following are some ideas that may impact a person's perspective about the length of a creation day:

- Holy Spirit inspired authors to write the Bible; they used the language and figures of speech of their culture
- A Western mindset does not always interpret a Hebrew figure of speech (*merismus*) correctly
- Hebrews associated the terms "evening" and "morning" only later as being twelve hours each to comprise a twenty-four-hour day
- God of the Bible does not need time to create since he lives in eternity
- Time was created on Day Four for the needs of people, not God

- Days of creation are stated as linear periods (day one, day two, etc.) because people can only think in a linear manner

Mystery: How much time was used for each day of creation?

## LENGTH OF TIME BETWEEN CREATION DAYS THEORY

Some creationists think the time between each creation day might have been longer than 24 hours, even if the creation day was 24 hours. They believe this allows time between days so the prior day's creations can mature before the next creation enters the environment. This would result in the entire creation process extending beyond a 24-hour creation week, that Young Earth Creationists believe.

## ISSUE SHOULD NOT BE LENGTH OF DAY

The issue for students of the Bible should not be how long a creation day was but what Jesus did on these days. For example, in the verses below, Jesus creates light where there is only darkness on the surface of the Earth.

> In the beginning God *(Jesus)* created the heavens and the earth. The earth was formless and void, and darkness was over the surface of the deep, and the Spirit of God was moving over the surface of the waters. Then God *(Jesus)* said, "Let there be light;" and there was light. God *(Jesus)* saw that the light was good; and God *(Jesus)* separated the light from the darkness. God *(Jesus)* called the light day, and the darkness He called night. And there was evening and there was morning, one day. (Genesis 1:1–5 NASB, author's emphasis)

## CHAPTER QUESTIONS

DQ1: Why do you think believing in a particular length of a creation day is unnecessary for believing in biblical creation?

DQ2: Why should the foundation of our Christian beliefs in biblical creation be based on what Jesus did on each creation day, not how long he took?

DQ3: Why does the Hebrew use of a merismus seem a logical way to understand the length of a creation day?

DQ4: Why is the amount of time Jesus used in creation a mystery?

## DIG DEEP

DD1: We can learn from theories, but we must always retain the essential biblical truths as the foundation of our beliefs. When have you not done this and been led astray by an idea or theory that later proved incorrect?

## PERSONAL APPLICATION

PA1: How has this study of the Hebrew word for a day helped clarify your understanding of the length of a creation day?

# 10.

## Delving Deeper: How Long Have People Been on Earth?

Important: Delving Deeper chapters contain material that may be difficult to understand. I suggest you spend extra time studying it and ask for the Holy Spirit's help. The Father and Jesus sent him to live within Christians as their Teacher, Guide, and Helper.

### BATTLEFRONT: AGE OF HUMANITY

One of the greatest battlefronts between those who believe in creation and evolution are theories about the age of humanity. However, evolution has always been and always will be a theory that cannot be proven. In contrast, the Bible is continually proven by science and nature, proving it is the truth and living word of God for everyone.

> When some people suggest the age of humanity, they also purport to know the age of the universe. Perhaps the ages of these do not coincide.

The following verses say everything in the Bible is true and useful for teaching because it is inspired by the God of the Bible. This means the creation account in Genesis is true and, therefore, trustworthy.

> All Scripture is inspired by God and is useful to teach us what is true and to make us realize what is wrong in our lives. It corrects us when we are wrong and teaches us to do what is right. God uses it to prepare and equip His people to do every good work. (2 Timothy 3:16–17 NLT)

# AGE OF HUMANITY

The following are theories about how long people have been on the Earth. There are other theories about the age of humanity in addition to these.

## Young Earth Creationist Biblical Genealogy Theory

Genealogies in the book of Genesis mention descendants from Adam to Abraham. They include the ages of the men. For example, Methuselah lived 969 years. Based on the assumption that no other direct descendants were missing in the genealogy lists, Young Earth Creationists theorize that Adam was created about 6,000 years ago. They propose 6,000 years is the age of humanity, the Earth, and the universe. (Some think they are up to 10,000 years old.)

## Old Earth Creationist Theory

They do not consider genealogies in their theories about creation. In contrast to Young Earth Creationists, they believe a creation day was a long period, not 24 hours. Some believe God may have created the universe 13.8 billion years ago and the Earth 4.5 billion years ago. They are not evolutionists.

## Population Growth Theory

One population growth model suggests that it would require 12,000 years for a global population to increase from 4 million to 10.4 billion by 2058. They estimate the global population in 10,000 BC was about 4 million. They don't estimate how long it took to reach the 4 million. No one knows for sure, so this is a guess based on assumed early population growth rates.

## Male Y-Chromosomal Theory

One Y-chromosome theory suggests people have been on Earth from 200,000 to 300,000 years. This is the "Y-chromosomal Most Recent Common Ancestor" (Y-MRCA) theory, informally known as the "Y-chromosomal Adam." The timeline is based on an assumed unbroken line of male ancestors for the Y-chromosomal-Adam that goes back to the first anatomically modern humans.

## Evolutionist Theories

According to biological evolution theories, modern human beings (Homo sapiens) evolved about six million years ago. They think we gradually developed our modern human traits and behaviors from prior ancient primates over millions of years. Another theory says that human bipedalism (walking on two legs) evolved about four million years ago. A recent human genome study theorizes that modern humans evolved from earlier non-human ancestors about 250,000 to 300,000 years ago.

## COMPARISON OF THEORIES

These theories about how long people have lived on the Earth are widely different. Some have more science involved, while others have more biblical support. Which is true?

Following is a comparison of these theories:

- Young Earth Creationist: 6,000 years (some say up to 10,000 years)
- Old Earth Creationist: 3.8 billion years
- Population Growth (to reach 10.4 billion in 2058): 12,000 years
- Male Y-Chromosomal: 200,000 to 300,000 years
- Evolution Genome Study: 250,000 to 300,000 years
- Evolution (walking on 4 legs): about 4 million years
- Evolution (ancient origins): about 6 million years

Each of these theories has its proponents from science, Christians, and the general population. Most contradict the other theories, so not all of them can be true.

## CHAPTER QUESTIONS

DQ1: Why do you think the length of time humans have been on the Earth is so controversial?

DQ2: What's the greatest weakness in the population growth theory?

DQ3: What's the greatest weakness of the Male Y-chromosomal theory?

DQ4: Which of the age of humanity theories seems most logical to explain how long people have been on Earth?

## DIG DEEP

DD1: What is the reason some people can have a different view of the age of the universe and the age of humanity? In other words, why are they based on the same logic and information?

## PERSONAL APPLICATION

PA1: Why do you find a particular theory more believable than the others?

# 11.

# Delving Deeper: Creation Myths from Other Cultures

Important: Delving Deeper chapters contain material that may be difficult to understand. I suggest you spend extra time studying it and ask for the Holy Spirit's help. The Father and Jesus sent him to live within Christians as their Teacher, Guide, and Helper.

## CREATION MYTHS

Under the influence of the Holy Spirit, Moses wrote the book of Genesis at a time when surrounding cultures had creation stories. The existence of these other extrabiblical stories about creation helps to reinforce the reality of the biblical creation account. Several minor myths about creation exist from other cultures, but Enuma Elish, the Babylonian Creation Myth, is the most widely known.

### Enuma Elish (The Babylonian Creation Myth)

This myth is also called *The Seven Tablets of Creation*. The tablets were discovered in 1849 in the ancient Royal Library of Ashurbanipal in modern-day Mosul, Iraq. It was written somewhere between the eighteenth and twelfth centuries BC. (Some scholars believe Moses wrote the book of Genesis around the fifteenth century BC while the Israelites wandered in the wilderness.) The Enuma Elish mythological story has been compared to the biblical account of creation due to a few similarities. However, the stories are significantly different in critical areas.

Summary of the Enuma Elish myth: It tells the story of the Babylonian god Marduk's victory over the chaos and violence brought about by the lesser gods. Creation is only incidental in this poem of praise for Marduk. Instead, it espouses the great feats of the Babylonian god Marduk. The myth says there are many gods where magic is the source of power. These gods are subject to nature and magic. People were created to serve these gods.

Critical differences between the creation account and this myth are as follows:

- Genesis is based on the Bible and is not a myth, while Enuma Elish is a myth.
- Genesis' God is monotheistic (one God), while the Enuma Elish is polytheistic (many gods).
- Genesis states that the Triune God is not created and exists from eternity, while the gods of Enuma Elish are created beings.
- Genesis's account emphasizes creation, while Enuma Elish is a poem or hymn to praise the god Marduk (patron deity of Babylon) as the head over the other gods; creation is incidental to his story.
- Genesis's creation is one of orderliness and peace, while Enuma Elish is one of chaos, violence, and warfare.
- Genesis's creation account includes everything that exists, while creation in Enuma Elish excludes vegetation, animals, the sun, and light.
- Genesis's God holds all power over nature and all creation, and he is subject to no one or nothing, while with Enuma Elish, magic spells are the source of power, so the gods are subject to nature and magic.
- Genesis's people were created in God's image and were given authority over the natural realm, while with Enuma Elish, people were created to serve the gods according to their will.

Website: More information about this myth can be found on the following website: https://carm.org/other-questions/does-the-genesis-creation-account-come-from-the-babylonian-enuma-elish/

## PHILOSOPHY OF RELIGION

The following is a true story about how a study of religion can lead to a wrong path to the God of the Bible:

> In undergraduate school in college, I took a class in philosophy that focused on world religions. We studied the major philosophers and their views on religion. We studied Christianity along with other world religions. At the end of the class, the professor asked what religion did we want to believe in: Christianity, which says people are bad and sinful, or another religion that says people are good. The professor didn't teach us that the God of the Bible is the only living God who loves people and wants to have an eternal relationship with them.

After reading this book, you will know what this professor didn't understand: that the Christian God loves people so much that he sent his only Son to die on the cross to pay the penalty for all their sins (John 3:16). When you accept Jesus as your Savior, his resurrection from the dead provides what no other religion can: eternal life with the Triune God and a changed life with a new nature like Jesus.

## MAJOR WORLD RELIGIONS AND CREATION

All major world religions (except Buddhism) believe in the existence of one or more gods. You can learn more details about their beliefs by searching the Internet. I will briefly review Christianity, Judaism, Hinduism, Buddhism, Islam and their views on creation and God.

---

All major world religions (except Buddhism) believe in the existence of at least one god. Yet they are very different in their beliefs about God himself and creation.

---

### Christianity

In 2024, there were 8.1 billion people worldwide, and about 2.4 billion people declared themselves to be Christian. This makes it the largest religion by population in the world. The three largest Christian populations are: Catholics (50 percent), Protestants (37 percent), Eastern Orthodox Church members (12 percent), and 1 percent as other. Christianity is named after the New Testament Savior and Lord Jesus Christ. It began as a religion 2,000 years ago after his death, resurrection, ascension, and the outpouring of his Spirit at Pentecost. It started with just over 2,000 converts from Judaism. Jesus was Jewish and followed the Mosaic Law while he was on Earth, so Christianity's roots are in Judaism.

It is a monotheistic religion, believing in one true God, which is the Trinity of God as the Father God, his Son, Jesus Christ, and the Holy Spirit. The Trinity means one God in three unique and equal persons of God. Unlike their Jewish spiritual ancestors, Christians believe that Jesus Christ is the long-awaited Messiah of Israel. Salvation is a free gift of God received by grace and faith that Jesus died and rose again as the eternal Savior and Lord. Christians read and study the Old and the New Testaments of the Bible.

Creation. Traditional Christians accept the biblical creation account in Genesis chapters 1 and 2. They believe the Creator God (Trinity) is the Father God as the Intelligent Designer, Jesus is the one speaking everything into existence, and the Holy Spirit is the one who has breathed life into people. Many believe the universe and Earth were created in six literal 24-hour days. However, some think a creation day may have been longer than 24 hours. Some Christians believe the universe had an Intelligent Designer (Father God) who used Jesus to create and guide it to evolve over 13.7 billion years (as evolutionists believe). This does not infer they believe in the theory of evolution.

### Judaism

In 2024, there were about 16 million Jewish people worldwide. Many, however, do not profess Judaism as their formal religion. A little over 40 percent of Jewish people live in Israel, and about the same percentage live in the United States. Judaism is small in population compared to the other major world religions, but it is the center of God's plan for the redemption of the world. Many believe that Judaism began when the Father God called

Abram out of his hometown of UR in the land of the Chaldeans and renamed him Abraham. You can read about the story of Abraham in Genesis chapters 11 through 17.

Judaism is a monotheistic religion where people believe there is only one person of God, who is referred to in the Old Testament of the Bible by many names and titles, such as Elohim, Jehovah, and Father. They do not believe in the triune nature of God the Father, Jesus Christ, and the Holy Spirit, as do Christians. Their religious books are the Tanakh (Old Testament to Christians) and the Talmud (oral commentary by various rabbis). They do not believe that Jesus Christ is the long-awaited Messiah of Israel. Therefore, they do not read and study the New Testament. Salvation is a matter of keeping the Law of Moses to the best of a person's ability and living a life of faith in the Father God.

Creation. Orthodox Jewish people accept the Genesis account of creation as true but see Jehovah, not the Triune God, as the one who created everything. The ultra-Ortodox Jews accept the creation account and reject the theory of evolution. In contrast, other Jewish people reject the biblical account of creation because it contradicts the theory of evolution. Most Jewish people are not religious. Rough estimates are that only about 2 to 3 percent of Jewish people in Israel are born-again Christians. Only these would accept that Jesus was the Jewish Messiah and co-author of creation.

**Islam**

In 2024, Islam was the world's second-largest religion, with an estimated 1.9 billion followers called Muslims. They are the majority population in 49 countries of the world. Islam is a monotheistic religion. They believe in one god, Allah. They do not believe in the Trinity of the Father God, his Son, Jesus Christ, and the Holy Spirit as Christians do. They believe Muhammad to be the last and most perfect Prophet of Allah. They consider Adam, Abraham, Moses, Jesus, and others in the Bible as additional human messengers of Allah. The Qur'ān (Koran) contains their sacred scriptures they believe were given to Muhammad by revelation.

Creation. The creation story from Islam can sound deceptively similar to the creation account in the Bible. Just like the God of the Bible, the god of Islam, Allah, existed for eternity. He first created his throne, which existed over the waters of everything. He then made the heavens, Earth, and everything in them in six days. Accordingly, Allah is seen as omnipotent in power and sovereign in authority. He simply spoke, and things came into existence out of nothing. He made all living creatures, angels, planets, and rain to make vegetation grow on the Earth.

The Muslim creation myth sounds biblical. But, it is not the creation account from Genesis chapters 1 and 2 of the Bible. This becomes clear when you understand the biblical creation account and that the God of the Bible is not the god of Islam. There is only one Creator God, the Trinity of the Bible, not Allah.

## Hinduism

In 2024, there were about 1.2 billion Hindu believers in the world. This makes it the third largest religion in the world, after Christianity and Islam. About 1.1 billion live in India. Most religious scholars believe it started between 2,300 BC and 1,500 BC in the Indus Valley near modern-day Pakistan. However, many Hindus believe their religion had no beginning and always existed. There is no founder to Hinduism, unlike Buddhism. Instead, it appears to be a compilation of many related beliefs over time. Hinduism is polytheism. They believe in many gods. However, some Hindus believe in one god with 33 different facets.

Creation. They believe that for an indeterminate period, only chaos and water existed everywhere. The supernatural being, Lord Vishnu, appeared floating on a lotus flower. The flower grew from his navel and split into three parts to form the heavens, the Earth, and the sky. Out of his loneliness, he split himself into two parts, male and female. All people were created from these two. This, however, is only one mythological story about creation in the Hindu religion.

Reincarnation. Hindus believe in reincarnation. They believe people exist in a continuous cycle of death and rebirth called "Samsara." When a person's physical body dies, their soul is reborn into a different physical body. If they do more good in a reincarnated life, they receive good "karma," and if they do more bad, they receive bad karma. Karma carries over into the next reincarnated life and affects that life. The final stage for the soul after many reincarnations is "Nirvana." This is a spiritual state of bliss, free from all human desires.

## Buddhism

In 2024, there were an estimated 500 million followers of Buddhism worldwide. About 50 percent live in China. Buddhism and Hinduism have common origins in ancient India. Siddhartha Gautama, the founder of Buddhism, lived from 563 to 483 BC in Nepal. The myth is that one day, he fell into a deep state of meditation under a bodhi (fig) tree, the "tree of wisdom." He achieved the highest state of god-consciousness, called Nirvana. He spent the next 40 years teaching the wisdom of the truths he learned. He became known as the "Buddha" or "Enlightened One."

There is no god identified to worship in Buddhism. Therefore, some religious scholars don't consider it a religion. Instead, they consider Buddhism a philosophy of life with moral codes of conduct. I asked an Asian friend I grew up with what he thought of Buddhism. He said he didn't consider it a religion since it focused on moral behavior and not on a god.

Creation. They don't believe in the initial creation of the world. Instead, they believe the world has been recreating itself millions of times. Buddhists don't believe the universe had a specific beginning. As a result, they see no need for a creator god. They believe in supernatural beings who help or hinder people from reaching "enlightenment." Some believe in reincarnation and say that after death, they can be reborn into another life on Earth. Others believe in a "Buddhist Creation Myth." This myth explains how people became bound to the "wheel of Samsara" and life after life in the "Six Realms."

Important: You can research these world religions on the Internet if you want to learn more about their beliefs. Remember, though, they are not biblically grounded religions and, as such, will not lead to truth and a personal relationship with the God of the Bible. Only Christianity can do this.

## CHAPTER QUESTIONS

DQ1: The Enuma Elish (The Babylonian Creation Myth) has likely existed for 3,500 years. What makes it such an enduring myth about creation?

DQ2: Why should Christians know what the other major world religions believe about creation?

DQ3: How can this help their faith in the God of the Bible and the one true world religion of Christianity?

DQ4: Which of the creation myths of these major world religions seems furthest from the biblical creation account? And why?

## DIG DEEP

DD1: Why do billions of people believe in a religion without experiential proof of its validity?

## PERSONAL APPLICATION

PA1: In which god do you want to believe? Your god? Someone else's god? Or the God of the Bible? And why?

PA2: How would you respond to a Muslim or Buddhist who asked you why you believed in the God of the Bible?

# PART D

Creation of the Earth

# 12.

## Creation Out of Nothing

### ONLY THE CREATOR GOD CAN MAKE SOMETHING OUT OF NOTHING

In the following verses, the God of the Bible says he is the Creator, and only he can create something out of nothing.

> Where were you when I laid the foundations of the earth? Tell Me, if you know so much. Who determined its dimensions and stretched out the surveying line? What supports its foundations, and who laid its cornerstone as the morning stars sang together and all the angels shouted for joy? Who kept the sea inside its boundaries as it burst from the womb, and as I clothed it with clouds and wrapped it in thick darkness? For I locked it behind barred gates, limiting its shores. I said, "This far and no farther will you come. Here your proud waves must stop!" Have you ever commanded the morning to appear and caused the dawn to rise in the east? Have you made daylight spread to the ends of the earth. (Job 38:4–13 NLT)

When the Father God, the Intelligent Designer, decided to create everything, his Son, Jesus, spoke the universe and Earth into existence out of nothing. *Ex Nihilo* is a Latin phrase that means "out of nothing." The following verses help us to understand this:

> God stretches the northern sky over empty space and hangs the earth on nothing. (Job 26:7 NLT)

> By faith we understand that the entire universe was formed at God's command, that what we now see did not come from anything that can be seen. (Hebrews 11:3 NLT)

> Mystery: In the beginning, nothing existed, so how did Jesus create something out of nothing?

## CREATING THE NATURAL ELEMENTS: PERIODIC TABLE

The creation account reveals the Triune God's majesty, unique power, and all-encompassing authority to design and create something out of nothing.

### First Substances Created

I believe that before Jesus created anything, he first created the natural elements he would use for the entire creation. (This is an opinion based on my understanding of Scripture and chemistry.)

### The Periodic Table

Today, we can identify the known natural elements in chemistry's Periodic Table. There are currently 118 elements in the Table. However, only the first 92 elements naturally occur, so I believe Jesus created them. Elements 93 through 98 (neptunium, plutonium, americium, curium, berkelium, and californium) occur naturally in tiny amounts due to human activities (for example, nuclear fallout) or naturally occurring radioactive decay. These six elements are found in uranium-rich pitchblende. Therefore, some periodic tables include them as naturally occurring. However, since the six did not likely exist at the creation of the universe and the Earth, I exclude them as having been created by Jesus.

> Jesus Christ created all the natural elements out of nothing. This included the Earth. Once he did this, he created every living plant and animal, including people, out of these natural elements.

Elements 93 through 118 are sometimes called "synthetic elements." Jesus did not create them. Scientists artificially created these by manipulating the fundamental particles of existing natural elements in a nuclear reactor, a particle accelerator, or by the explosion of an atomic bomb. No element with an atomic number greater than 99 has any value other than for scientific research.

Instructional Comment: Elements in the Periodic Table are listed numerically by their assigned "atomic number" (1 to 118). They are identified as solids, liquids, and gases.

Website: The Periodic Table with 118 elements can be found on the following website: https://sciencenotes.org/periodic-table-wallpaper-118-elements

### Elements, Atoms, and Molecules

A comparison of these is an advanced topic that I will simplify with some basic definitions.

Definition: *Elements* were created by Jesus or later synthetically made by scientists. They are the most basic substances on Earth or in space. They consist of only one atom or a certain type of atom.

*Atoms* are the smallest amount of an element.

*Molecules* are the result of combining two or more atoms of elements. For example, water is the molecule H2O.

### Molecule Example: Water

When Jesus created water, he combined two atoms of the Hydrogen element with one atom of the Oxygen element. This combination resulted in a "molecule" of water. In chemistry, we refer to water by its molecular composition, "H2O."

Mystery: How did Jesus know how to create molecules for everything that exists?

## NATURAL ELEMENTS ON EARTH

The following verses indicate that Jesus made people, animals, and birds from natural elements he had already created. These elements were part of the "ground" that was made when Jesus created the Earth out of nothing (Genesis 1:1).

Then the LORD God formed the man from the dust of the ground. (Genesis 2:7 NLT)

So the LORD God formed from the ground all the wild animals and all the birds of the sky. (Genesis 2:19 NLT)

## EARTH'S CRUST

The Earth's crust is the outer layer of our planet where plant, animal, and sea life reside. Beneath it is the mantle and core, which is molten lava. On average, the crust is about 25 miles deep. It is composed of solid rocks and minerals, which, theoretically, result from the cooling and solidification at the Earth's birth. Scientists can find all 92 naturally occurring elements in the crust at various places and levels of the earth's surface.

Following are the ten most common natural elements found in the Earth's crust (by weight):

- Oxygen (46.1 percent)
- Silicon (28.2 percent)
- Aluminum (8.23 percent)
- Iron 5.63 percent)
- Calcium (4.15 percent)
- Sodium (2.36 percent)

- Magnesium (2.33 percent)
- Potassium (2.09 percent)
- Titanium (0.565 percent)
- Hydrogen (0.140 percent)

## NATURAL ELEMENTS IN SPACE

I believe the Father God designed everything to be made from the natural elements. Therefore, when Jesus created the universe and all its cosmic objects, he did so using natural elements. Since the universe was created out of nothing, Jesus would have made the 92 natural elements, and then he created the universe from these.

Astronomers and astrophysicists theorize that the vastness of the universe is mostly empty, a near-perfect vacuum. Evidence from observations with telescopes, mathematical calculations, and experimentations corroborates what we know from the Bible, that the 92 naturally occurring elements of the Periodic Table are found in space. Of course, when we say they occur in space, we include those found on planet Earth.

Following are the ten most common natural elements found in space (by mass):

- Hydrogen (70 percent)
- Helium (28 percent)
- Oxygen, Carbon, Neon, Nitrogen, Magnesium, Silicon, Iron, and Sulfur

Note: You will learn in the chapter, "Creation of the Universe" that solid matter (such as planets and stars) composes only 5 percent of space. The 95 percent that is not visible solid matter is theorized to be dark matter and dark energy.

## ANGELS NOT CREATED FROM NATURAL ELEMENTS

You have already read that angels are supernatural beings that were created before the creation of the universe and Earth. Since they are supernatural, they cannot have been made from the natural elements in the Periodic Table. The devil and his demons were angels before their rebellion, so they are made out of the same supernatural substance as the angels of God.

Mystery: What kind of substance did Jesus create to make angels?

## GENETIC TEMPLATES FROM NATURAL ELEMENTS

Our Father God is innovative in his designs for creation. It seems he took the natural elements to fashion genetic templates for making similar creation kinds. An article suggests that intelligent design is behind similarities in animal deoxyribonucleic acid (DNA). These similarities reveal genetic templates for organisms with the same anatomical and functional components. For example, these templates can be seen in the genetics of animals for such things as bones, muscles, and organs. The above is from the article "The Untold Story Behind DNA Similarity," "Science in Perspective" in the *Answers, Building a Biblical Worldview* (answersmagazine.com, May, June, 2017).

Definition: *Creation kinds* were created by Jesus rather than individual species of organisms we see today. I see these kinds as "master species" that carry all the genetic material needed for generations of reproduction, leading to a variety of species as they reproduce over time.

## CHAPTER QUESTIONS

DQ1: What would you say to someone to help them understand what it means for Jesus to have created the universe and Earth out of nothing?

DQ2: How would you describe what "nothing" is?

DQ3: Why do you think the Father designed the universe to be mostly empty of solid matter?

DQ4: Why were angels created from a substance different than those natural elements in the Periodic Table?

## DIG DEEP

DD1: Explain why the Periodic Table must represent the natural elements created by Jesus out of nothing.

## PERSONAL APPLICATION

PA1: What have you learned about the Periodic Table and the natural elements of which your body is composed?

# 13.

# First, Second, Third, and Fourth Days

## DAYS OF CREATION OUTLINE

The seven days of creation are outlined below. Notice what is created each day.

| Day | Verses | Creation | Comment |
|---|---|---|---|
| 1 | 1:3–5 | Light; day and night | Separated light from darkness |
| 2 | 1:6–8 | Sky | Separated waters from heavens and Earth |
| 3 | 1:9–13 | Seas, dry ground, and vegetation | Vegetation for food |
| 4 | 1:14–19 | Sun, moon, stars; seasons, years, and days | Our solar system created; light shines on Earth; time created |
| 5 | 1:20–23 | Sea life and birds of the air | Includes sea and aerial dinosaurs |
| 6 | 1:24–31 | Domestic & wild land creatures; People | Includes land dinosaurs; People given rulership over creation |
| 7 | 2:1–3 | Creation finished | Day of rest; Seventh day is holy |

Table 1: Seven days of creation

## FOUR DAYS TO PREPARE THE EARTH

Genesis 1:1 says, *In the beginning God created the heavens and the earth* (NET). But when was the "beginning?" Scripture does not tell us when Jesus created the universe and the Earth. It seems at some point after its creation, the Earth was "without shape and empty." *Now the earth was without shape and empty, and darkness was over the surface of the watery deep, but the Spirit of God was moving over the surface of the water* (Genesis 1:2 NET). We see that Jesus used creation days One, Two, part of Three, and Four to modify the Earth from a useless place for life to one where life could thrive.

## FIRST DAY: LIGHT CREATED

Light was created, and then it was separated from the darkness from which it was taken. It doesn't say the light cast illumination onto the Earth's surface. It says the purpose of the light was to create a separation of daytime and nighttime. But where did the light come from?

> Then God said, "Let there be light;" and there was light. God saw that the light was good; and God separated the light from the darkness. God called the light day, and the darkness He called night. And there was evening and there was morning, one day. (Genesis 1:3–5 NASB, author's emphasis)

Bible commentator Albert Barnes[1] says the following about the creation of light on the First Day:

> *The first day's work is the calling of light into being.* Here the design is evidently to remove one of the defects mentioned in the preceding verse—"and darkness was upon the face of the deep." The scene of this creative act is therefore coincident with that of the darkness it is intended to displace. The interference of supernatural power to cause the presence of light in this region, intimates that the powers of nature were inadequate to this effect. But it does not determine whether or not light had already existed elsewhere, and had even at one time penetrated into this now darkened region, and was still prevailing in the other realms of space beyond the face of the deep. Nor does it determine whether by a change of the polar axis, by the rarefaction of the gaseous medium above, or by what other means, light was made to visit this region of the globe with its agreeable and quickening influences. We only read that it did not then illuminate the deep of waters, and that by the potent word of God *(Jesus)* it was then summoned into being. *This is an act of creative power, for it is a calling into existence what had previously no existence in that place*, and was not owing to the mere development of nature. Hence, the act of omnipotence here recorded is not at variance with the existence of light among the elements of that universe of nature, the absolute creation of which is affirmed in the first verse. (Albert Barne's Notes on the Bible, author's emphasis)

Information: Albert Barnes (1798–1870) was an American theologian, pastor, and author. He is known for his extensive Bible commentary on the Old and New Testaments of the Bible. The "Albert Barnes Notes on the Bible" is thorough and accurate for Old and New Testament Bible passages.

## SECOND DAY: SKY CREATED

On the second day, the sky above the Earth was created. This is the place where we see birds fly. Sometimes, it is referred to as the atmosphere around the Earth.

> Then God said, "Let there be an expanse in the midst of the waters, and let it separate the waters from the waters." God made the expanse, and separated the waters which were below the expanse from the waters which were above the expanse; and

---

1. Barnes, *Notes*, Genesis 1:3–5.

it was so. God called the expanse heaven. And there was evening and there was morning, a second day. (Genesis 1:6–8 NASB)

Instructional Comment: When Jesus created the sky, he moved the clouds with water vapors to be above the sky and the waters on the Earth to be below the sky. Birds don't fly above the clouds in the Earth's upper atmosphere. They fly in the expanse (sky) between the clouds and the Earth. The sky is sometimes referred to as the "heavens" in the Bible, as in Genesis 1:9.

## THIRD DAY: SEAS, DRY LAND, AND VEGETATION CREATED

On the third day, dry land and seas (oceans) were created. Both were designed to be suitable for living creatures to flourish.

> And God said, "Let the waters under the heavens be gathered together into one place, and let the dry land appear." And it was so. God called the dry land Earth, and the waters that were gathered together he called Seas. And God saw that it was good. (Genesis 1:9–10 ESV)

Note: Initially, Jesus moved all the waters "into one place." I believe he would ultimately spread the waters out into our saltwater seas, oceans, and freshwater lakes and rivers.

### Dry Land and Water

We know from these verses that Jesus moved the waters on the surface of the Earth to create seas/oceans and cause dry land to appear. Today, about 71 percent of the Earth's surface is covered in water. About 95 percent of this water is found in our saltwater seas/oceans.

Following are some amazing facts about how much water is on Earth:

- If the Earth's surface was flat, there would be enough water to cover it completely 1.8 miles deep.
- Lake Superior has enough water to cover North and South America in one foot of water.

## VEGETATION CREATED FOR FOOD

The Father designed land animals, including people, to eat only vegetation for food. He prepared the Earth on Day Three with the food they needed to thrive for their eventual creation on Day Six.

> Then God said, "Let the land sprout with vegetation—every sort of seed-bearing plant, and trees that grow seed-bearing fruit. These seeds will then produce the kinds of plants and trees from which they came." And that is what happened. The land produced vegetation—all sorts of seed-bearing plants, and trees with

seed-bearing fruit. Their seeds produced plants and trees of the same kind. And God saw that it was good. (Genesis 1:11–12 NLT)

### Father's Imaginative Designs

The Father's unimaginable (to us) imagination can be seen in his creation of a myriad of varied plant life. Modern scientists have discovered nearly 400,000 plant species. New species are constantly being discovered. Some are delicate in their branches, stems, and flowers, while others are strong, thick, and hardy. Some are incredible and unique in their artistic design and practical function. They all received their nourishment from the soil and sun. All were eternal, and nothing died.

### Corruption of Plant Life

But things changed for the entire creation after the fall of Adam and Eve. When the resultant corruption entered the world, all plant and animal life experienced the sentence of physical death. Plant life changed. It dies, rots, and perishes in the ground. Among the most uniquely changed plants is the "Venus Flytrap," which became carnivorous. It lives by eating insects and small animals such as mice and frogs. Its "mouth" is actually a modified, hinged leaf. Some of the leaves are colorful, which helps attract prey. Some secret a sweet scent that also attracts prey. There are over 600 known species of the Venus Flytrap.

## CREATION KINDS

In Genesis, we see that everything living (plant and animal life) was made using "creation kinds." I suspect that a creation kind may have been a "master species." By this, I mean that each had a comprehensive and diverse genetic composition that allowed them over time to develop, form new variations, and adapt to different locations and environmental conditions.

> Jesus brought into existence creation kinds when he made plant and animal life. Perhaps human beings were also a creation kind that includes Neanderthals. That is one theory about their existence.

### Hybridization

This is the process by which two parents from different varieties or species within the same creation kind produce offspring. The offspring can display physical traits distinct from those of either parent. Hybridization is used in many areas of biology and botany today. For example, it is used to produce new varieties of roses.

If two modern creatures (perhaps from different species) can hybridize with true fertilization and create offspring, these two creatures are descended from the same biblical creation kind. The opposite may not be true. If two creatures cannot hybridize, they may not necessarily be from different kinds. This is because the loss and deterioration of genetic material over time may have prevented the hybridization, even though they were from the same kind.

### Designed Intelligence to Reproduce

The Father God designed all seed-bearing plants with a genetic intelligence to reproduce from their seeds according to their creation kind. In the chapter, "Delving Deeper: Genetics Validates Creation," you will learn how he designed cells with the DNA needed for each species to function as he intended.

> Mystery: How does a plant seed know what it is to reproduce? For example, should it produce an apple tree or a cucumber?

## FOURTH DAY: SUN, MOON, AND STARS CREATED

Jesus created vegetation on Day Three for food, but without the sun's illumination and warmth on the Earth, it would not grow.

> Then God said, "Let lights appear in the sky to separate the day from the night. Let them be signs to mark the seasons, days, and years. Let these lights in the sky shine down on the earth." *And that is what happened. God made two great lights—the larger one to govern the day, and the smaller one to govern the night. He also made the stars.* God set these lights in the sky to light the earth, to govern the day and night, and to separate the light from the darkness. And God saw that it was good. And evening passed and morning came, marking the fourth day. (Genesis 1:14–19 NASB, author's emphasis)

### Theories about the Illumination of the Earth's Surface

One theory says that the sun, moon, stars (universe), and Earth already existed as a result of their creation identified in Genesis 1:1. It theorizes that light didn't reach the Earth's surface because of a dense, dark atmosphere around the Earth (Genesis 1:2). Therefore, only diffused light became visible on Day One. Its source (the sun) was not yet visible on the Earth's surface. The theory says that on Day Four, Jesus cleared up the Earth's atmosphere so that the sun, moon, and stars could be seen, and the sun's illumination and warmth could reach the Earth's surface: *God placed the lights in the expanse of the sky to shine on the earth* (Genesis 1:17 NET).

A second theory is that Genesis 1:1 is a summary of creation, which means Jesus created the sun, moon, and stars (universe and Earth) on Day Four of creation. Neither theory suggests the universe was created on Day Four. As with all these theories, trying to understand the verses literally and sequentially is difficult.

## FOURTH DAY: TIME CREATED

In Genesis 1:14–15, we read:

> And God said, "Let there be lights in the expanse of the heavens to separate the day from the night. And let them be for signs and for seasons, and for days and years, and let them be lights in the expanse of the heavens to give light upon the earth." (ESV)

The creation of the sun and moon provides a method for people to calculate time.

*Seasons*, Hebrew is *mô'êd*; an appointment, a fixed time

*Years*, Hebrew is *shâneh*; a year, as a revolution of time

*Day*, Hebrew is *yôm*; has many applications for time in the Bible

Albert Barnes[2] says the following about the creation of time:

> While the first day refers only to the day and its twofold division (evening & morning), the fourth (day) refers to signs, seasons, days, and years. These lights are for "signs." They are to serve as the great natural chronometer (time) of man, having its three units—the day, the month, and the year—and marking the divisions of time. (Albert Barnes' Notes on the Bible)

### Time Not Needed by God

God lives in eternity where there is no time. He does not need time to help him manage the "passing of time." We see in the following verse that time is irrelevant to him:

> But you must not forget this one thing, dear friends: A day is like a thousand years to the Lord, and a thousand years is like a day. (2 Peter 3:8 NLT)

### Sun and Moon "Rule Over" Day and Night

We read in the following verses that the sun and moon were given to help people experience the periods known as daytime and nighttime:

> And God made the two great lights—the greater light to rule the day and the lesser light to rule the night—and the stars. And God set them in the expanse of the heavens to give light on the earth, to rule over the day and over the night, and to separate the light from the darkness. (Genesis 1:16–16 ESV)

---

2. Barnes, *Notes*, Genesis 1:14–15.

## CHAPTER QUESTIONS

DQ1: The Earth's surface is 71 percent water (mostly salt water). What are the ways people use this much water in their lives?

DQ2: Why do you think the first theory about the illumination of the Earth might be a reasonable explanation to clarify the differences in Genesis 1:1–17?

DQ3: How does hybridization demonstrate the reality of creation kinds?

DQ4: Why do you think Jesus waited to create time until creation Day Four since he uses the idea of time for each prior Creation Day?

## DIG DEEP

DD1: Explain how the seed of a tomato vine knows to produce tomatoes, not potatoes.

## PERSONAL APPLICATION

PA1: What would your life be like if you didn't have time to manage your daily activities?

# 14.

# Fifth and Sixth Days: Life in Seas and on Earth

### FIFTH DAY: SEA LIFE CREATED

All sea life was created on Day Five. This would have included life in saltwater and freshwater.

> Then God said, "Let the waters teem with swarms of living creatures, and let birds fly above the earth in the open expanse of the heavens." God created the great sea monsters and every living creature that moves, with which the waters swarmed after their kind, and every winged bird after its kind; and God saw that it was good. God blessed them, saying, "Be fruitful and multiply, and fill the waters in the seas, and let birds multiply on the earth." There was evening and there was morning, a fifth day. (Genesis 1:20–23 NASB)

## Oceans and Seas

Oceans are saltwater and cover about 71 percent of the Earth's surface. Some examples are the Pacific Ocean, Atlantic Ocean, and Indian Ocean. Seas are also saltwater, smaller in geographic size, and usually land-enclosed and linked to an ocean. There are 76 seas in the world. Some examples are the Black Sea, the Red Sea, and the Philippine Sea. Saltwater is about 3.5 percent salinity (sodium chloride) and various dissolved ions.

Definition: *Sea life* is all living creatures existing in saltwater oceans and seas.
   *Aquatic life* includes all sea life and freshwater life.

## Diversity of Sea Life

Our oceans and seas are filled with an incredible diversity of sea life. These creatures thrive in saltwater environments.

Categories of sea life are:

- Fish (including eels and sharks)
- Amphibians (few species tolerate saltwater)
- Invertebrates (for example, clams, coral, kelp, and squid)
- Mammals (for example, whales, otters, and seals)
- Reptiles (for example, sea snakes, crocodiles, and various turtles, including dinosaurs)

Instructional Comment: Tiny organisms, called plankton, provide food for the largest sea creature, the whale. In comparison, fish are the preferred food source for sharks.

## FreshWaters

Freshwater accounts for about 3 percent of the Earth's surface water. Bodies of freshwater do not contain minerals that create saltwater. From ponds to huge lakes and many rivers, they exist globally and are the foundation of all plant, animal, and human life. *Without freshwater, nothing survives on Earth.*

Lake Superior is the largest freshwater lake in the world by surface area. Its surface area is 37,600 square miles ( approximately the size of Austria), with a maximum depth of 1,332 feet. The Amazon River is the second longest river in the world (4,000 miles), slightly behind the Nile River (4,258 miles). However, it is the largest in the world in terms of volume and width. The Amazon River is a massive, intricate water system weaving through one of the world's most vital and complex ecosystems, the Amazon rainforest in South America. The river and its basin are home to many unique species of animals, trees, and plants only found in its region.

## Diversity of Freshwater Life

Freshwater life consists of a diversity of aquatic life that parallels saltwater life.

Categories of freshwater life are:

- Fish
- Amphibians (for example, frogs, toads, and salamanders)
- Invertebrates (for example, clams and snails)
- Mammals (for example, beavers, otters, and river dolphins)
- Reptiles (for example, alligators, snakes, and various turtles)

These include various sizes, shapes, textures, and feeding habits. Just as living creatures from the oceans and seas provide food for the world's nations, so does freshwater life. In fact, look at a world map and see how many cities have been built adjacent to an ocean, sea, or freshwater river or lake. This is because people settle near their food supply. Billions of people globally rely on our oceans alone for their daily food supply, with many others relying on freshwater food.

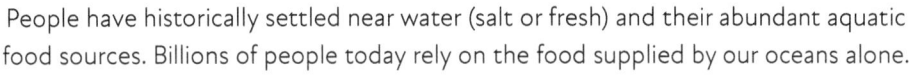

People have historically settled near water (salt or fresh) and their abundant aquatic food sources. Billions of people today rely on the food supplied by our oceans alone.

Note: Jesus created many types of birds, some living near saltwater (such as seagulls) and others that live near freshwater (such as ducks).

## FIFTH DAY: BIRD LIFE (AVIAN) CREATED

The Father God designed a fantastic variety of bird life to populate the Earth. Some are aquatic, while others thrive in pasture lands and cities. Some eat fish, while others eat insects or plant seeds. Each type of bird reproduces according to its creation kind. Birds are highly adaptable and are found in diverse global climates, from cold to hot tropical rainforests. Some, like the American Bald Eagle, soar in the sky while others sit peacefully in the trees. There are over 10,000 species of birds in the Aves Family. That's where we get the term avian for bird life.

Following are their top five distinctive avian characteristics:

- Warm-blooded: They maintain their body temperature internally rather than relying on the external environment to help regulate it. This allows them to live successfully in extreme cold and hot temperatures.
- Respiration Rate: Birds have a high respiration rate that enables them to take in large amounts of oxygen for flight.
- Feathers: Most bird species use wings and feathers for flight, warmth, and courtship. Some species, such as the ostrich, have short wings and feathers but cannot fly.
- Beaks and No Teeth: All birds have a beak or bill but no teeth. Food is swallowed and ground for digestion in the gizzard.
- Eggs: All birds produce hard-shelled eggs for reproduction. The fertilized embryo grows within this protective environment.

Consider the creative imagination of the Father God, who designed the beautiful Rainbow Lorikeet, the ugly vulture, and the Hoatzin, nicknamed the "stinkbird." For those who are patient, bird-watching can be a fascinating adventure of avian color and bird behaviors.

Note: Flying reptiles (avian dinosaurs) were designed by the Father and created through Jesus Christ. You will learn more about dinosaurs in the next chapter.

## SIXTH DAY: LAND LIFE CREATED

Every living thing on the land of the Earth was created on Day Six.

> Then God said, "Let the earth bring forth living creatures after their kind: *cattle and creeping things and beasts of the earth after their kind;*" and it was so. God made the beasts of the earth after their kind, and the cattle after their kind, and everything that creeps on the ground after its kind; and God saw that it was good. (Genesis 1:24–25 NASB, author's emphasis)

### Diversity of Land Life

Again, we can see the creative design of the Father God in land life. For example, Jesus created massive land animals like dinosaurs and elephants and tiny insects like ants.

Genesis 1:24–25 identifies three types of life created to live on the land:

- Cattle or livestock domesticated to live with people (for example, cows, horses, and goats)
- Beasts of the earth that live in the wild (for example, lions, wolves, and bears)
- Creeping things are smaller animals that move in a prostate posture on or off the ground (for example, insects and spiders)

Categories of land life are:

- Amphibians that live in or near water (for example, frogs and salamanders)
- Invertebrates (for example, insects, spiders, and worms)
- Mammals (for example, cows and wolves)
- Reptiles (for example, snakes and turtles, as well as dinosaurs)

Definition: *Mammals* are warm-blooded vertebrates that have mammary glands to breastfeed their young.

*Animals* include mammals, amphibians, reptiles, birds, and insects. They can be vertebrates (mammals, reptiles, and birds) or invertebrates (insects) and move under their own power.

## ADAM NAMED EVERY LIVING CREATURE ON THE EARTH AND SKY

According to Genesis 2:19–20, Adam named every creature living on the ground and the birds of the air (before Eve was created). Since I believe dinosaurs were made on creation days Five and Six, Adam would have named these as well.

> The LORD God formed out of the ground every living animal of the field and every bird of the air. He brought them to the man to see what he would name them, and whatever the man called each living creature, that was its name. So *the man named all the animals, the birds of the air, and the living creatures of the field*, but for Adam no companion who corresponded to him was found. (Genesis 2:19–20 NET, author's emphasis)

## There Were Many Creatures to Be Named

There were a lot of creatures for Adam to name. We don't know how many were created, and we don't know how many exist today. One estimation indicates there may be nearly two million species of living creatures on Earth today. However, nobody knows for sure how many species of everything exist. The range of estimations is wide among scientists who study species. It spans from a few million to billions.

> Mystery: How long did it take Adam to name all the living creatures? Are the names Adam made up the same names we use today?

The mysteries of creation can enhance our imagination and draw us closer to the Creator God.

## CHAPTER QUESTIONS

DQ1: Name one creature for each category of sea life.

DQ2: Why are feathers essential for warm-blooded bird life?

DQ3: Why were all land-dwelling animals vegetarians when created?

DQ4: Why is it a mystery about how long it took Adam to name all living creatures?

## DIG DEEP

DD1: Explain why amphibians and mammals needed to be designed differently.

## PERSONAL APPLICATION

PA1: What would life be like for you if all living creatures did not have names?

# 15.

## Fifth and Sixth Days: Dinosaurs

The word dinosaur means "terrible lizard." However, they are not lizards. They are an extinct and separate group of reptiles. Some scientists say the only true dinosaurs were land dinosaurs and that aquatic and avian dinosaurs were reptiles and not dinosaurs. While others say they are all dinosaurs. I refer to them all as dinosaurs in this book.

### FIFTH DAY: AQUATIC AND AVIAN DINOSAURS CREATED

When all life in the seas was created on Day Five, Jesus made aquatic dinosaurs according to their creation kinds.

> Then God said, "Let the waters teem with swarms of living creatures, and let *birds fly* above the earth in the open expanse of the heavens." God created the *great sea monsters* and every living creature that moves, with which the waters swarmed after their *kind*, and every winged bird after its kind; and God saw that it was good. God blessed them, saying, "Be fruitful and multiply, and fill the waters in the seas, and let birds multiply on the earth." There was evening and there was morning, a fifth day. (Genesis 1:20–23 NASB, author's emphasis)
>
> *Great sea monsters*, Hebrew for *great* refers to something large, while *monsters* is is *tannîyn* and refers to aquatic or land monsters; *Monsters* are considered sea serpents or jackals, dragons, serpents, or whales. (I include aquatic dinosaurs as well.)

Note: There is no direct reference in these verses to inform us that avian dinosaurs were created on Day Five. However, the fact that they existed leaves us to conclude they were created on Day Five along with birds.

### Dinosaur Creation Kinds

The Father designed, and Jesus called into existence dinosaurs by various creation kinds, just as he did all living creatures. Like all creation kinds, dinosaurs of the water, sky, and land had massive genetic material, enabling them to adapt over thousands of years. I see these kinds as "master species" that carry all the genetic material needed for generations of reproduction, leading to a variety of species as they reproduce according to the command of Jesus.

> Like all plant and animal life, dinosaurs were created as creation kinds. The massive amount of genetic material in their original creation kind allowed them to adapt and thrive globally in different climates.

### Imaginative Dinosaur Designs

I think dinosaurs were created as several creation kinds, perhaps as simple as water, sky, or land dwelling kinds. Or maybe more specific kinds unknown to us. Whatever their creation kinds were, the Father used his imagination to design an incredible diversity of dinosaurs. For example, the diversity includes the powerful carnivore, T-rex, huge herbivores, such as sauropods, "duck bill" dinosaurs, and small rodent-type dinosaurs. Some dinosaurs were bipeds (walked on two legs), while many others walked on four legs. Some were aquatic, living in water, while others flew or lived on land. All dinosaurs were cold-blooded reptiles.

### Common Design Templates

Earlier in this book, you learned about genetic templates the Father designed and Jesus used to create animals. The DNA found in cells provides animals of similar creation kinds with similar anatomical and functional characteristics. For example, they all have a muscular skeleton system primarily composed of bone and muscle. This idea of genetic templates applies to the creation kinds of dinosaurs.

Scientists classify dinosaurs by a common hip design (design template by the Father). There are two templates of hip design used across many dinosaur species. We can see hip design in two-legged and four-legged dinosaurs. Again, we see the Intelligent Designer and Jesus working together to create species according to the Father's detailed plan.

### Aquatic (Water) Dinosaurs

In the Bible, it seems that dinosaurs of seas are referred to as "Leviathan."

> Can you pull in Leviathan with a hook, and tie down its tongue with a rope? (Job 41:1 NET)

> In that day the LORD will punish Leviathan the fleeing serpent, With His fierce and great and mighty sword, Even Leviathan the twisted serpent; And He will kill the dragon who lives in the sea. (Isaiah 27:1 NASB)

> *Leviathan*, Hebrew is *livyâthân*: a wreathed animal, that is, a serpent (especially like the crocodile or some other large sea monster)

Aquatic dinosaurs are separated into four groups as follows:

- Plesiosaurs
- Pliosaurs
- Mosasaurs
- Ichthyosaurs

> The Hebrew words for aquatic dinosaurs imply some were "twisted" in their physical appearance.

The Plesiosaurs and Pliosaurs groups are examples of this, with their long necks that could appear twisted as they moved in the water.

A short list of some of the more terrifying flesh-eating aquatic dinosaurs follows:

- Spinosaurus (seen in Jurassic Park movie)
- Liopleurodon (exceptionally fast)
- Shastasaurus (size of a blue whale)
- Hydrorion (gigantic dinosaur with a long neck)
- Dakosaurus (crocodile-looking dinosaur)
- Archelon (prehistoric sea turtle; largest turtle to exist)

Instructional comment: There are approximately five hundred known species of aquatic dinosaurs. All but one were carnivores eating shellfish, crustaceans, and fish. The larger ones ate smaller aquatic reptiles.

Website: The following website provides detailed descriptions with pictures of aquatic dinosaurs from an evolutionist's perspective: https://www.americanoceans.org/facts/water-dinosaurs/

## Freshwater Aquatic Dinosaurs

Fossilized bones from an aquatic dinosaur that apparently lived in freshwater were recently found in an ancient riverbed system in Morocco. The bones were from several "Kem Kem Plesiosaurs." These were previously considered to be marine (sea-going) dinosurs only. Their fossils indicate they routinely lived and fed near the shore and in freshwater, along with fish, frogs, turtles, and crocodiles. Paleontologists don't know why some of these marine dinosaurs were in freshwater.

## Non-Feathered Avian Dinosaurs

If you have watched Sci-Fi movies about prehistoric times, you likely saw Pterodactyl dinosaurs flying in the sky. Fossils of these reptiles have been found to solidify their reality.

They belong to a subgroup of flying reptiles known as Pterosaurs and are theorized by evolutionists to have existed between 163.6 to 66 million years ago. These flying dinosaurs did not have feathers. Instead, they had skin tightly covering their bones. They were carnivores, mostly eating fish and small land animals. Unlike aquatic and land dinosaurs, which were cold-blooded, avian dinosaurs (like modern birds) were warm-blooded. Some scientists consider these to be flying reptiles and not dinosaurs.

A short list of known non-feathered avian dinosaurs from the scientific Order of Pterosauria follows:

- Pteranodon (known as the "flying toothless dinosaur")
- Pterodactylus (powerful flight capabilities)
- Dimorphodon (hang from tree branches and cliffs, holding on with its claws)
- Rhamphorhynchus (narrow jaws with sharp teeth that protrude outward)
- Germanodactylus (size of a raven with a head crest that would change color)
- Haopterus (Teeth adapted for eating fish, could travel long distances at high elevations)
- Quetzalcoatlus (one of largest flying dinosaurs with an average wingspan of 52 feet)

**Feathered Avian Dinosaurs**

Evolutionists theorize that feathered bird-like flying reptiles from the Theropod group are the ancient ancestors that evolved into modern birds. One primary factor is the existence of feathers on these flying reptiles. Avian dinosaurs from this group had flight feathers, whereas those from the Pterosauria Order did not. Like the Pterosauria avian dinosaurs, these also were carnivores.

Website: The following website provides detailed descriptions with pictures of avian dinosaurs from an evolutionist's perspective: https://dinosaurfactsforkids.com/list-of-flying-dinosaur-names-all-pterosaur-species/

Note: We know, as creationists, that evolution is a theory that cannot be proven. Instead, we know that Jesus created all dinosaurs for the water, sky, and land.

## SIXTH DAY: LAND DINOSAURS CREATED

All land animals were created on Day Six, so this would have included land-based dinosaurs.
    Then God said, "Let the earth bring forth living creatures after their kind: cattle and creeping things and beasts of the earth after their kind;" and it was so. God made the beasts of the earth after their kind, and the cattle after their kind, and everything that creeps on the ground after its kind; and God saw that it was good. (Genesis 1:24–25 NASB)

One internet site says there are 42 known land dinosaurs. Most of these would be unfamiliar unless you study dinosaurs. The movie series Jurassic Park brought some to life on the big screen. If you remember the movies, some were docile plant eaters (herbivores), while others were vicious meat eaters (carnivores). About 65 percent of land dinosaurs were herbivores, and 35 percent were carnivores (or omnivores).

Following are some of the better-known land dinosaurs:

- Allosaurus (one of largest predators with lengths around 40 feet, carnivorous)
- Brontosaurus ("thunder lizard," easily recognizable due to its massive size, long neck, and tail, herbivorous)
- Diplodocus (82 feet tall, long neck and tail, herbivorous)
- Parasaurolophus ("duck-billed dinosaur," herbivorous)
- Protoceratops (herbivorous)
- Stegosaurus (curious looking with verticle plates along back spine, favorite meal for the ferocious allosaurus, herbivorous)
- Triceratops ("three-horned face," distinctive three horns and upward shield behind head, herbivorous)
- Tyrannosaurus (one of the largest land-based meat-eaters, strong, fast, intelligent, fast, and excellent sight and smell, carnivorous)

Website: The following website provides detailed descriptions with pictures of land dinosaurs from an evolutionist's perspective: https://www.activewild.com/list-of-dinosaurs-names-with-pictures/#dinosaur-index

## Dinosaurs Initially Plant Eaters

All animal life was initially herbivores, that is, plant eaters. This included all species of dinosaurs. There was no killing for food or any other reason. Only after Adam and Eve's fall, when the Earth was corrupted, did dinosaurs begin to kill for food.

# CREATIONIST AND EVOLUTIONIST DINOSAUR THEORIES

How did dinosaurs really come into existence? Did they evolve as evolutionists propose, or did Jesus create them? There is a vast amount of physical proof that dinosaurs roamed the Earth, waters, and sky millions of years ago. So, let's review these two perspectives on their origin.

## Creationists Dinosaur Perspective

Those who believe in creation believe that all animal life was created during the Fifth and Sixth days of creation. This included all creatures that evolutionists refer to as prehistoric.

Creationists believe the Earth is thousands of years old, not hundreds of millions or billions of years old, as evolutionists do. I believe dinosaurs were amazingly designed by the Father God and created by Jesus on the Fifth and Sixth Days of creation.

### Dinosaur Information and Speculation (Creation)

The following speculation about dinosaurs is based on Young Earth Creationism:

- First appeared on the Earth at creation about 6,000 years ago
- Went extinct about 4,000 years ago (flood or ice age dates)
- Number of dinosaur species is estimated to be 1,000 or more (Dr. Clarey)

### Creationist Book on Dinosaurs

Dr. Tim Clarey, a Young Earth Creationist, wrote a detailed and illustrated book, *The Science of the Biblical Account, Dinosaurs, Marvels of God's Design*, that explains the creation of dinosaurs. He provides significant scientific and biblical evidence that supports creation and debunks the theory that dinosaurs evolved. He includes information from creation and evolutionary scientists throughout the book. His book covers such topics as the creation of dinosaurs, their life on Earth, recent dinosaur discoveries, the Father God's design of dinosaurs, dinosaur anatomy, biology, and behavior, and their extinction.

### Evolutionist Dinosaur Perspective

Evolutionists theorize that dinosaurs first appeared on the Earth between 243 and 233 million years ago. However, some evolutionists disagree about how long ago dinosaurs lived, over what period, and how they became extinct. They know they existed, but they don't know how long ago.

### Dinosaur Information and Speculation (Evolution)

The following is speculation by evolutionary scientists. All dates and numbers are theories.

- First appeared on the Earth between 243 and 233 million years ago
- Went extinct about 66 million years ago
- Number of dinosaur genera (plural for genus) is estimated to be 1,000 or more

## DINOSAUR SOCIAL BEHAVIOR

Dr. Clarey discusses dinosaur behaviors. He is quick to state that these are speculations based on fossil evidence that cannot be proven. Dinosaurs are reptiles and lay eggs instead of giving live birth as mammals do. He says their eggs and nests provide evidence that some dinosaurs were social. For example, some females lay their eggs in circular nests along with other females of the same species. One scientist theorizes that the spacing between the nests seems to indicate they distanced their nests from each other enough so they did not step on and destroy the eggs of the others.

### Instinct or Learned Behavior?

How do we get into the mind of an extinct dinosaur? Well, of course, we cannot. But, we can extrapolate from the evidence of their behaviors. Considering the above nesting evidence, Dr. Clarey suggests it represents social behavior. But is it instinctive or learned? There is an important difference. All animals, including dinosaurs, have instinctive behaviors. Instinct means they don't think first and then react. They simply react to the specific situation. Learned behavior is the result of experiences in life. They learn it is for their benefit to do (or not do) something. They learn to avoid suffering loss or to do something for positive gain. Did these dinosaurs learn to avoid stepping on the eggs of others to avoid retaliation, or was this an instinct they just knew to do?

### Being Social

What human behaviors and characteristics come to mind when you think of someone as "social?" How do they behave around others? Are they polite, gracious, considerate, engaging, complimentary, and so on?

### T-Rex Social Behavior

I recently watched a show on the National Geographic channel entitled "Drain the Oceans." Season Six, Episode Three, was about a recently discovered new species of dinosaurs theorized to be 65 million years old. They stated that many species of dinosaurs had existed. For example, there are multiple species of Tyranosaurs. Yes, not all "T-Rex" are like the ones we see in the movies. In studying the fossils remains of multiple Tyranosaurs found in one spot in Utah and others found in Montana and Alberta, Canada, they discovered that these large killing machines hunted in packs like wolves do. And that they had social groups and likely families that lived and hunted together. This conclusion changed the way scientists think about the Tyranosaurs species.

Important: Even though followers of Christ do not believe in the theory of the evolution of species (macroevolution), we can learn about God's creations from scientific discoveries.

## DINOSAUR AND HUMAN COEXISTENCE CONTROVERSARY

As creationists, we believe all dinosaurs were created on Days Five and Six. This means they began their existence thousands of years ago along with all living creatures, including people. The GenesisPark.com website says its primary purpose is to provide evidence that dinosaurs existed and coexisted with people over a lengthy period. Its website has much scientific evidence, including pictures and articles donated by many individuals, Christian organizations, and museums. Many of these help validate the claim that dinosaurs existed simultaneously with humans. They provide reports and pictures of dinosaurs and human footprints together, demonstrating that they coexisted. They identify ancient footprints found globally in places like Glen Rose, Texas (Willet Print, Feminine Print, Delk Track); New Mexico (Zapata Track); Utah (Meister Print); Tuba City, Arizona; Kentucky; Tanzania (Laetoli Track); and Turkmen Republic.

Website: More information about the creationists' beliefs on this coexistence can be found on the following website: https://www.genesispark.com/exhibits/evidence/paleontological/footprints/

Note: Evolutionists attempt to discredit this evidence with various explanations to purport that dinosaurs lived hundreds of millions of years *before* the earliest human ancestors. They theorize that our early human ancestors (Homo erectus and Homo habilis) came into existence about ½ million years ago, not 243 million years ago when they say dinosaurs first appeared. Therefore, they theorize dinosaurs and people could not have coexisted.

## CHAPTER QUESTIONS

DQ1: Why do you think the Father God may have delighted in his imaginative designs for the diversity of dinosaurs?

DQ2: Describe the difference between feathered and non-feathered flying dinosaurs.

DQ3: Which type of dinosaur (avian, aquatic, or land) would have been most feared by people living concurrently with them? Why?

DQ4: Personification is applying human characteristics to non-human things. Why are dinosaur social behaviors not personifications?

## DIG DEEP

DD1: How would creationists and evolutionists use the empirical evidence from nature differently to validate their theories about dinosaurs and human coexistence?

## PERSONAL APPLICATION

PA1: In what ways does dinosaur social behavior reflect your social behavior?

# 16.

# Dinosaur Extinction and Fossils

## DINOSAUR EXTINCTION THEORIES

Some creationists believe that dinosaurs were not on the ark or they were on it but died out shortly after the flood. Either way, they believe dinosaurs went extinct several thousand years ago. Evolutionists believe that an extraordinary catastrophic event occurred that wiped dinosaurs out about 66 million years ago.

## TWO EVOLUTIONIST EXTINCTION THEORIES

Neither evolutionist nor creationist theories can be proven since no one was there when dinosaurs went extinct. The following are two prominent evolutionist theories about how and when dinosaurs went extinct.

### Meteor Impact Theory

The first theory is that a massive meteor crashed into the Earth, filling the atmosphere with vast amounts of poisonous gases, dust, and debris about 66 million years ago. The result was a major alteration in the climate of the Earth. This caused pollution in the atmosphere, which also blocked critically needed sunlight. As a result, the Earth's warm, humid surface grew colder. The gasses poisoned plant and animal life, and the dust settled and suffocated living creatures. The lack of sufficient healthy plant life resulted in the demise of plant-eating dinosaurs.

### Volcano Eruption Theory

The second theory is that a massive volcano in India erupted, sending huge amounts of lava, poisonous gasses, and debris into the atmosphere. This is known as the "Deccan

Traps." Scientists who support this theory estimate it occurred about 66 million years ago. They say the volcanic lava covered about 200,000 square miles of the Earth's surface up to 6,000 feet deep. Like the theorized meteor impact, this also would have altered the Earth's climate and had similar catastrophic impacts, causing the extinction of dinosaurs along with many other animal and plant life.

## THREE CREATIONIST EXTINCTION THEORIES

Creationists believe dinosaurs were created with all other animals on creation Days Five and Six. Therefore, they would have existed during the pre-flood period. There is no indication in Scripture that they became extinct before then. These theories posited by many creationists are based on this pre-flood assumption.

### Noah's Ark Theory

Dinosaurs would have been safely aboard the ark before the flood (unless God chose not to save them). Perhaps they were young adults, the larger possibly being the size of a cow.

> With them in the boat were *pairs of every kind of animal*—domestic and wild, large and small—along with birds of every kind. Two by two they came into the boat, *representing every living thing that breathes*. A male and female of each kind entered, just as God had commanded Noah. Then the LORD closed the door behind them. (Genesis 7:14-16 NLT, author's emphasis)

### Reduced Dinosaur Population Theory

Creationists theorize the world after the flood was much different than before. For one thing, they believe the climate was cooler. Cold-blooded reptiles would have more difficulty surviving in colder climates. Also, dinosaurs and other predator animals would have hunted and killed other dinosaurs and animals. Therefore, it's likely their global population was drastically reduced after the flood. Some theorize people may have contributed to hunting the remaining dinosaurs to extinction.

### Ice Age Theory

Many Creationists believe the Ice Age started as a result of cataclysmic volcanic eruptions that sent vast amounts of ash and gases into the atmosphere. Others believe the Ice Age was started as a result of a large meteor crash that sent massive amounts of dust into the atmosphere. As a result, the reduced sunshine, high humidity, and cold after the flood resulted in massive, deep ice formations. And this is what created the "Ice Age."

### Estimated Ice Age and Flood Dates

Following are estimated dates for the Ice Age by Young Earth Creationists, who believe it occurred shortly after the flood. They base these estimates on the belief that the heavens and Earth were created about 6,000 years ago.

Theorized Ice Age and flood dates by Young Earth Creationists follow:

- Global Flood (Noah) started about 2,348 BC
- Ice Age began shortly after the flood, about 2,250 BC, and ended about 2,000 BC

Evolutionists theorize the Ice Age started about 2 million years ago and ended about 11,000 years ago. Notice the vast differences between the creationist and evolutionist estimates:

- When the Ice Age started and ended
- How long it lasted

## FOSSILS AND FOSSILIZATION

There is a lack of understanding of what fossils actually are and how they form. The following should help to resolve both issues.

### What's a Fossil?

Have you ever picked up a fossilized bone? If so, you noticed it was much heavier than it would have been as a living bone. This is because the living tissues died and were "mineralized." Over time, the tissues decayed and were replaced with minerals in the ground. Most of the time, only the hard tissues, such as bone and shells, survive long enough to become a fossil. Since bones and shells have pores, the pores are filled with minerals. Thus, completely mineralized fossils are solid rocks.

Some of the common minerals found in many bone fossils are as follows:

- Calcium carbonate (one atom of calcium and carbon, three atoms of oxygen, $CaCO_3$)
- Iron
- Silica
- Calcite (one atom of calcium, carbon, and oxygen, $CaCO$)

Fossils give us a glimpse into the past, a view of what plant and animal life was like thousands of years ago. Dinosaur fossils can trigger our imaginations of what life might have been like if we lived amid these enormous plant eaters and voracious carnivores like the mighty T-rex.

### What Is Fossilization?

Fossilization is the process by which the soft tissues (muscles, organs, and so on) and hard tissues (bones) of an animal are preserved over time. Only in very rare environments can soft tissue be preserved for thousands of years—for example, the La Brea Tar Pits, solid ice glaciers of Siberia, and deep in the oceans where little to no oxygen exists.

Paleontologists estimate that less than 1 percent of all plant and animal species that ever lived have been fossilized. They believe that less than 1 percent of all dinosaur species that ever lived have been found.

There are several explanations for this rarity of fossils:

- Most animals die and decay or are eaten by other animals above ground
- Conditions for fossilization are rare
- Many fossils are buried so deep in the earth that they are never seen

The best conditions for fossil formation are as follows:

- Rapid burial (avoids scavengers and decay above ground)
- Lack of oxygen (oxygen expedites decay, so long-term mineralization cannot occur)
- High mineral content is needed in the ground to create fossils

Instructional Comment: Riverbeds, lakes, and oceans provide the best conditions for fossilization.

Website: The following evolutionary website provides details about how fossilization occurs: https://prehistoricsaurus.com/dinosaur-fossils/fossilization-process/how-do-fossils-form/

Reminder: Evolutionists believe dinosaur fossils are hundreds of millions of years old, while creationists believe they are thousands of years old.

### How Much Time Is Required for Fossilization?

One website states that at least 10,000 years were required for fossilization. They said this is because fossils are defined as the physical remains or traces of plants and animals that died over 10,000 years ago. Therefore, the minimum time for something to become a fossil is 10,000 years. But this really doesn't answer the question.

A better answer. The amount of time for fossilization depends upon the environment in which the remains are preserved. Some estimated 20,000-year-old dinosaur remains have been found with soft tissue and little mineralization. As creationists, we know that dinosaurs lived on Earth thousands, not millions of years ago. Therefore, the age of fossils can be estimated in thousands, not millions of years. But how many thousands of years?
Website: More information about fossils can be found on the following Young Earth Creationists creation website: https://creation.com/a-fossil-is-a-fossil-is-a-fossil-right

## RADIOMETRIC DATING

Scientists, especially geologists, use radiometric dating to estimate the age of rocks and the fossils contained within them. This is also called radioisotope dating or radioactive dating. Only unstable natural elements decay and generate isotopes that can be measured. For example, in radiocarbon dating, Carbon-12 and Carbon-13 are stable and do not decay. Only Carbon-14 is unstable and decays. This is complicated, but the unstable atoms decay into stable "daughter atoms" at a stable rate. Scientists estimate the age of a rock by measuring the number of its unstable atoms and comparing it to the number of stable daughter atoms in the rock. This result is the amount of time since the rock was initially formed. Mineralized dinosaur fossils can be roughly dated in this way.

### New AI-Based Dating Method

Standard dating methods using radiocarbon dating have proven to be unreliable at times. New research to find more accurate methods of dating fossils has resulted in AI-based dating of genomes. This new AI technology enables DNA to be analyzed more accurately to date human fossil remains up to ten thousand years old. Temporal Population Structure (TPS) is used to date human genomes. It is not intended to replace radiocarbon dating. Instead, it is seen as a complementary method for dating human fossil remains when radiocarbon dating may not be sufficiently accurate.

Website: More information about AI-based dating can be found in the following article: https://www.sciencedaily.com/releases/2022/08/220823162730.htm

Note: More advanced methods of dating the Earth and fossils are needed. Perhaps an advanced combination of genetic research and technology (especially AI) will provide more accurate and reliable dating results.

## NO DINOSAUR DNA READABLE AFTER 10,000 YEARS

Interestingly, many evolutionary scientists don't believe DNA can be found in dinosaurs since they theorize they lived more than 68 million years ago. They know DNA doesn't last that long, so they are skeptical about reports that readable Dinosaur DNA has been found. (Readable means short sequences of the DNA are visible in fossil remains and can be compared to live tissue.)

### Fragile DNA Example

Evolutionary scientists recently found a Tyrannosaurus rex (T-rex) dinosaur estimated by evolutionary standards to be 65 to 68 million years old. Yet, it still had soft tissues, blood vessels, and blood cells. Fragile DNA was extracted from its bones. For the DNA to be readable, the bones had to be thousands of years old, not millions. Also, DNA can only be extracted if the remains have not completely fossilized. Again, scientists have validated creation and disproved evolution.

---

Worldview: Dead dinosaurs don't lie. The existence of readable dinosaur DNA validates creation and disproves evolution.

---

One study on Moa bird dinosaur bones revealed that DNA has a half-life of 521 years. This means that by 10,000 years, no DNA molecules in dinosaur bones can be read. This result assumes optimal environmental conditions for preservation.

Website: The following Institute of Creation Research website helps explain why DNA has a half-life of 521 years from the Moa dinosaur study results: https://www.icr.org/article/dna-dinosaur-bones

## Why Is This Important to Creationists?

Once again, scientists and evolutionists validate creation and disprove evolution. If dinosaurs are supposed to have died out about 66 million years ago and dinosaur DNA only exists for thousands of years, then evolutionists cannot validate their claims about the evolution of dinosaur species.

## CHAPTER QUESTIONS

DQ1: "Dead dinosaurs don't lie." What does DNA have to do with the accuracy of this statement?

DQ2: Why should Christians learn about the existence and extinction of dinosaurs on Earth?

DQ3: Why are fossil bones heavier than similar non-fossilized bones?

DQ4: What are some problems with the accuracy of dating fossils?

## DIG DEEP

DD1: What occurs if fossils don't have optimal environmental conditions for preservation? How can a careless analysis of a discovery like this lead to an erroneous conclusion about the age of the fossil?

## PERSONAL APPLICATION

PA1: Why is the extinction of dinosaurs not an issue for your belief in the God of the Bible?

# 17.

## Sixth Day: Ancient Human Ancestors?

Creationists believe that people and all living creatures on land were created on Day Six.

> God said, "Let the land produce living creatures according to their kinds: cattle, creeping things, and wild animals, each according to its kind." ... Then God said, "Let us make humankind in our image, after our likeness, so they may rule over the fish of the sea and the birds of the air, over the cattle, and over all the earth, and over all the creatures that move on the earth." (Genesis 1:24–26 NET)

Note: As you will see in this chapter, I don't believe extinct hypothetical human ancestors ever existed.

### EVOLUTION'S ANCIENT HUMAN ANCESTORS

Evolutionists propose that we have ancient human ancestors who existed millions of years ago.

According to their classification, all are bipeds, while some are hominins.

Definition: *Bipedal*: A species classified as bipedal by evolutionists walks on two legs and shares common anatomical characteristics with modern people. Therefore, some primates and all humans are bipedal:

- Pelvis is short and wide
- Legs are longer than the arms
- Thigh bones slant inwards
- Hips are wide apart and knees are close together
- Thigh and shin bones join at an angle

*Hominin*: Evolutionists use the term hominin to represent modern humans (Homo sapiens), Neanderthals, erect bipedal primates, and extinct species of human-like ancestors (such as Homo erectus). These include species of the genus Homo and the great apes (orangutans, gorillas, chimpanzees, and bonobos). All hominins are also bipeds.

Website: You can learn more about bipedalism from the following website: https://australian.museum/learn/science/human-evolution/walking-on-two-legs-bipedalism/

Website: There is more about hominins on the following website: https://www.sapiens.org/teaching-unit/hominins/

Instructional Comment: Evolutionists theorize that apes started to walk on two legs about four or five million years ago and hominins about 1.8 million years ago. Of course, creationists disagree with this theory.

## EVOLUTION'S HUMAN LINEAGE THEORY

The human lineage theory is based on direct descent. This means that one species evolves into another over time. It requires the existence of "transitional Species."

Definition: *Direct descent*, according to the theory of evolution, means that one species evolves into another species.
*Transitional species* are the missing links needed to demonstrate that a prior species evolved into another new species.
Important: No transitional species can be proven to have existed in the human lineage theory.

The following is evolution's direct descent human lineage theory, including the type of species:

| Australopithecus afarensis | Homo habilis | Homo erectus | Homo neanderthalensis | Homo sapiens |
|---|---|---|---|---|
| *Ape-like biped* | *Human-like hominin* | *Human-like hominin* | *Human-like hominin or modern human?* | *Modern humans (hominin)* |

Table 2: Theorized human lineage chart with type of species

Note: More about Homo neanderthalensis in the next chapter, "Sixth Day: Neanderthals."

### Out of Africa Model: Homo Erectus

Some evolutionists theorize that Homo sapiens evolved from Homo erectus in Africa. They say these later migrated to different parts of the world. This is the "Out of Africa Model."

Adriana Heguy, a genomics researcher, says, "To my knowledge, no DNA had been extracted from Homo erectus fossils."

The author of this book could not find any scientific evidence that viable/readable DNA has been extracted from any proposed ancient human ancestor other than Neanderthals.

Instructional Comment: Since no Homo erectus DNA has been discovered, it is not possible to validate they were part of a human lineage theory. Therefore, it is impossible to prove the Out of Africa Model.

## FAKE AND MISTAKES IN HUMAN LINEAGE

Over the last hundred years, there have been several mistakes and at least one fake discovery of a purported ancient ancestor of modern humans. Some discoverers admitted they unwittingly made an initial mistake and later rectified it by acknowledging the fossil was not a human ancestor.

Following is a list of discoveries that creationists and many evolutionists now consider a fake or a mistake. However, some evolutionists steadfastly believe some of these mistakes are ancient human ancestors and missing links to modern humans. I include what the fossil was determined to be and the name of one of the scientists who disagreed and stated the discovery was not an ancient ancestor of modern humans.

- *Ramapithecus* (mistake): Ancient ape ancestor to an orangutan (Peter Andrews, "Hominoid Evolution")
- *Australopithecus* (mistake): Extinct form of an ape (Evolutionist Dr. Charles Oxnard, Professor of Anatomy and Human Biology at the University of Western Australia; many other evolutionists and scientists agree these fossils belong to an extinct ape; Charles Oxnard, *Fossils, Teeth, and Sex: New Perspectives on Human Evolution*)
- *Sinathropus*, Peking Man (mistake): Skull of an ape reconstructed by combining fragments of human skulls from surrounding areas (Evolutionists Marcellin Boule and Henri Vallois thought it might have been an ape killed by humans; Patrick O'Connell, *The Science of Today and the Problems of Genesis*)
- *Homo habilis*, Lucy (mistake): Although many evolutionists still believe that Lucy is an ancient ancestor to modern humans, some believe she was an ancient variety of a pigmy chimpanzee (Albert W. Mehlert, "Homo Habilis Dethroned," *Contrast: The Creation Evolution Controversy*)
- *Pithecanthropus erectus*, Java Man (mistake): Originally purported to be an ancient missing link, some 30 years later, scientists think it was an ancient ape killed by humans to eat its brain in Java (C. Loring Brace and Ashley Montegu, *Human Evolution: An Introduction to Biological Anthropology*)
- *Hesperopithecus haroldcookii*, Nebraska Man (mistake): First purported ancient human ancestor found in America was based on a tooth that was later proved to be from a pig (William K. Gregory, "*Hesperopithecus* Apparently Not an Ape nor a Man")

- Piltdown Man (fake): An intentionally damaged jaw of an ape with filled teeth was easily identified as a fraud (Malcolm Bowden, *Ape-Men: Fact or Fallacy*)
- *Homo neanderthalensis* (mistake): Homo neanderthalensis were purported to be the predecessor species to modern humans (Homo sapiens), but DNA evidence says otherwise. Multiple fossil discoveries and DNA prove beyond a shadow of a doubt that Neanderthals were not an ancient human ancestor. They lived and interbred with modern humans. Many evolutionists today classify Neanderthals as a subspecies of modern humans, Homo sapiens neanderthalensis (Boyce Rensberger and Jay Matternes, "Facing the Past")

Paul S. Taylor wrote *The Illustrated Origins Answer Book, Concise, Easy-to-Understand Facts about the Origin of Life, Man, and the Cosmos.* His book contains detailed scientific information about the origins of the universe, life, species, and mankind. The drawings and pictures are an excellent way to visualize the topics.

### Scopes Monkey Trial

In March 1925, John T. Scopes, a substitute Biology teacher in Tennessee, was put on trial for teaching Darwin's theory of evolution. The Butler Act was passed in 1925 by the Tennessee legislature. This new law made it illegal to teach in Tennessee schools about any theory contrary to the Bible's creation account in the Book of Genesis. After several trials, Scopes was indicted by a grand jury for unlawfully and willfully violating the Butler Act by teaching evolution. This law remained for more than four decades until the 1960s when the theory of the evolution of species was accepted in teaching in Tenessee schools.

Website: More information about the Scopes Monkey Trial can be found on the following website: https://www.history.com/news/90-years-ago-scopes-and-evolution-indicted-in-tennessee

## PARADIGM SHIFT FOR DIRECT DESCENT THEORY

The direct-descent model is a straight-line lineage from one species to another. The advent of genetics has caused modern evolutionists difficulty proving the validity of a one-to-one lineage. Due to newly discovered complexities in genetic compositions, they have problems defining species. As a result, some have abandoned direct descent with its related transitional species ("missing link") theory. This is not just a change but a classical "paradigm shift" in their thinking, research, and proposals.

Definition: *Paradigm Shift* is a radical change in the fundamental concepts that define something. The old ways of thinking and doing become irrelevant and are replaced by new ways.

Can you think of paradigm shifts in your own life? How did these affect your life positively or negatively?

# CHAPTER QUESTIONS

DQ1: Why do you think most evolutionists are atheists who have decided not to believe in the God of the Bible?

DQ2: Which entry in evolution's human lineage chart has been proven by science to be invalid and, therefore, supports creation?

DQ3: Why are evolutionists eager to prove evolution when they find new bones they attribute to human ancestors?

DQ4: Why has modern genetics created a paradigm shift for evolutionists, moving them away from direct descent human lineage theories?

# DIG DEEP

DD1: Explain why DNA must be "readable" to determine how old the fossil is.

# PERSONAL APPLICATION

PA1: Why does this list of frauds and mistakes about human evolution cause you to believe in creation and disbelieve in evolution?

PA2: How would you respond to an evolutionist asking why you thought the human lineage chart was false?

# 18.

## Sixth Day: Neanderthals

### NEANDERTHALS NOT TRANSITIONAL SPECIES TO HUMANS

A 2010 study of human DNA found that most people worldwide have between 1 and 4 percent Neanderthal DNA. This was the first study to provide solid evidence that modern humans interbred with Neanderthals. Their DNA is 99.7 percent identical to human DNA, while chimpanzees are 98.8 percent identical to humans. In other words, Neanderthals are closer genetically to modern humans than the closest primates (chimpanzees). Do you think people today can interbreed with chimpanzees? They cannot interbreed since they are separate species from different creation kinds. In another example, cats and dogs cannot interbreed and produce viable offspring because they are different species from different creation kinds.

Website: More information about Chimpanzee DNA can be found on the following website: https://answersingenesis.org/genetics/dna-similarities/untold-story-behind-dna-similarity/

Website: More information about this 1–4% Neanderthal DNA in modern humans can be found in the following Cornell University website article: https://news.cornell.edu/stories/2023/06/lingering-effects-neanderthal-dna-found-modern-humans

Note: My DNA profile from testing on a genealogy website showed I have Neanderthal DNA in this 1 to 4 percent range.

### NEANDERTHAL FOSSIL REMAINS AND DNA

Fossil remains of Neanderthals have been found globally. Their skeletal remains appear after the melting of the ice age. Since creationists believe the ice age occurred after the flood, Neanderthals might have been the offspring of Shem, Ham, or Japheth. They seemed to have lived in small groups and intermingled with modern humans at various times and places. The evidence implies there might not have been many of them.

Their fossil remains have been found in the following geographic areas:

- Engis, Belgium (first remains found in 1829; identified about 100 years later)
- Neander Valley, Germany (found in 1856; labeled Neanderthal from this valley; first published report of Neanderthal DNA)
- Fossilized skull from La Ferrassie, France
- Vindija Cave, Croatia (2008 study report; first mitochondrial DNA (mtDNA)) sequence for a Neanderthal)
- Mezmaiskaya Cave in Russia
- El Sidrón Cave in Spain
- Denisova Cave in Siberia
- Altai Mountains in Siberia

## Neanderthal Genes Help Modern Humans

The Wall Street Journal reported on two recent studies of modern human DNA (Nature and Science journals) that concluded we inherited genes from Neanderthals that helped us adapt to new environments. These were skin pigmentation, immune system function, and metabolism. This reflects the coexistence and interbreeding of Neanderthals and modern humans thousands of years ago, not millions.

## Neanderthal DNA

DNA has become an essential tool for scientists to analyze animal remains. However, DNA is fragile and degrades over short periods, so it only exists in more recent partially fossilized remains. It can help reconstruct the genome of purported ancestors of humans, such as Homo neanderthalensis.

Website: More information about this can be found on the following website: https://humanorigins.si.edu/evidence/genetics/ancient-dna-and-Neanderthals

Note: Partially fossilized remains are those that have not been completely mineralized into rock.

## Denisova Cave Study of Neanderthal and Human DNA

The initial results of this archaeological discovery date back to 2008. The study and subsequent reports involved multiple scientific disciplines. It was conducted on fossil remains in the Denisova Cave in Siberia. Included in the discoveries were fragments of ceramics, tools, and weapons. All of these, together with the DNA of the fossil remains, provided evidence that Homo sapiens, Neanderthals, and Denisovans lived in close proximity during an overlapping period. Therefore, *modern humans could not have descended from Neanderthals.* These scientists theorized the three were separate species.

One Christian author describing the Denisovan Cave discoveries stated that he believed these were three separate species of humans who were descendants of Adam and Eve. He believed they developed their somewhat different genetic and physical characteristics as a result of being spread all over the Earth by God at the Tower of Babel. However, he did not rule out the possibility that the three were one species descended from Adam and Eve.

Website: More about this Christian's perspectives can be read in his blog: https://newcreation.blog/how-many-human-species

As previously noted, studying the DNA of modern humans in various geographies found that Neanderthal DNA exists in many modern people groups. This also confirms that Neanderthals and modern humans lived and interbred in the same geographical areas. Here again, *science disproves the theory of human evolution. Neanderthals are not the transitional species to modern humans.*

---

Worldview: Dead Neanderthals don't lie. The existence of their readable DNA validates creation and disproves evolution.

---

Instructional Comment: If the three were indeed one species descended from Adam and Eve, they would have developed their genetic and physical differences through adaptation and microevolution.

## NEW EVIDENCE FOR THE HUMAN NATURE OF NEANDERTHALS

The following comes from a 2024 article in Evolution News that provides archaeological evidence of Neanderthal complex human behavior. It includes evidence from recent studies about their physical appearance, cognitive abilities, culture, and capabilities that reflect their human nature.

Note: Evolution News is a website that promotes theories about evolution. Christians, especially apologists, can use it as a source of current information about various aspects of the theories of evolution. As apologists, they can use biblical truth to counter the unprovable theories in discussions with evolutionists.

### Human-like Nature

Archaeological evidence from many discoveries tells us the following about Neanderthal human capabilities, creativity, and complex behaviors:

- Used fire (2023)
- Buried their dead (2020)
- Created stone circles (2016)
- Crafted tools from bones (2013)
- Crafted jewelry (2015, 2019)
- Made body decorations from feathers (2011, 2012)
- Created paintings and engravings as cave art (2014, 2018, 2023)
- Made bone flutes for music (2018)
- Made and used ochre for pigment (2012, 2018)
- Made sophisticated fibre products (2020)
- Produced plant-based flour (2023)
- Had a varied diet that included seafood (2020)
- Cooked their food; used herbs for painkillers and antibiotics (2012, 2017)
- Created a complex chemical process to make glue from birch bark (2021, 2023)

Website: The detailed report about Neanderthals can be found on the Evolution News website: https://evolutionnews.org/2024/02/fossil-friday-new-evidence-for-the-human-nature-of-Neanderthals

### More New Evidence

New evidence also suggests they were capable of complex communication similar to modern humans (2021). The study also cited that Neanderthals interbred with modern humans. This hybridization is only possible if they are genetically similar to modern humans (2023). The scientists behind the studies stated that modern humans possess 1 to 4 percent Neanderthal DNA, suggesting they share a common gene pool with modern humans.

Important: These various studies were conducted by a number of evolutionary scientists who wanted to understand who Neanderthals were and how they interacted with humans. Even though it was not their intention, *their conclusions support creation and disprove the theory of human evolution*. Once they remove Neanderthals as the transitional species to modern humans in the human lineage chart, they are left with a vacuum (missing link). No other ancient hominins can be proven to be the transitional species to humans. *The direct descent theory of the evolution of humans is dead.*

## WERE NEANDERTHALS CREATED IN GOD'S IMAGE?

The massive evidence that Neanderthals demonstrate the same nature as humans supports one of three theories about their origin:

- Theory One: Offspring of Adam and Eve (Same species, same creation kind) If they were the offspring of Adam and Eve, they may have been created in God's image.
- Theory Two: Not Offspring of Adam and Eve (Different species, same creation kind) If a different species, same creation kind, and not the offspring of Adam and Eve, they may not have been created in God's image (even though they interbred with Adam and Eve's offspring).
- Theory Three: Offspring of Shem, Ham, or Japheth (Same species, same creation kind) They would have been made in the image of God.

Note: So, how long have Neanderthals been on Earth? Since the beginning of time (creation), or since the flood.

## WHERE DID CAIN'S WIFE COME FROM?

Cain and Able were sons of Adam and Eve. After Cain killed his brother Able in a jealous rage, God cast him out of the land where Adam and Eve lived. Cain settled in the land of Nod (east of Eden), where he met and married his wife.

> Then Cain went away from the presence of the LORD and *settled in the land of Nod*, east of Eden. *Cain knew his wife*, and she conceived and bore Enoch. When he built a city, he called the name of the city after the name of his son, Enoch. (Genesis 4:16–17 ESV, author's emphasis)

> Mystery: How long were Cain and Able with Adam and Eve in peace before Cain killed Able?

Notice in the following verses how long Adam lived. Adam or his son, Seth, may have been the father of Cain's wife. Seth lived nine hundred and twelve years.

> When Adam had lived one hundred and thirty years, he became the father of a son in his own likeness, according to his image, and named him Seth. Then the days of Adam after he became the father of Seth were eight hundred years, and he had other sons and daughters. So all the days that Adam lived were nine hundred and thirty years, and he died. (Genesis 5:3–5 NASB)

Bible commentator John Gill[1] says the following about the potential wife of Cain:

> Who this woman was is not certain, nor whether it was his first wife or not; whether his sister, or one that descended from Adam by another of his sons, since this was about the one hundred and thirtieth year of the creation. At first indeed Cain could marry no other than his sister; but whether he married Abel's twin sister, or his own twin sister, is disputed. (John Gill's Expository of the Entire Bible)

---

1. Gill, *Expository*, Genesis 5:3–5.

Information: John Gill (1697–1771) was an English theologian, pastor, Bible scholar, and commentator. I find his Bible commentaries interesting and use them in my Bible studies.

### Who Was Cain's Wife?

Adam and Eve and their immediate descendants would have had a pure genetic composition without mutations. So it's genetically possible that Cain could have married one of his sisters, cousin, or niece and had genetically healthy children.

Instructional Comment: It was about 2,500 years later, after Adam and Eve, that God prohibited incest, which is sexual relations with a genetically close relative (Leviticus 18:6). Genetically close relatives today who marry could have children with genetic physical or mental abnormalities.

Another theory is that Cain's wife might have been a Neanderthal woman. This is a possibility since we know Neanderthals interbred with humans. Neanderthals were sufficiently genetically and biologically similar to humans to produce offspring, so this is not an impossibility. If so, was the Neanderthal woman made in God's image?

> Mystery: Who was the woman Cain married in the land of Nod? Was she a descendant of Adam and Eve or a Neanderthal woman?

Note: I am not sure Cain married a Neanderthal woman. It seems more likely that he married a descendant from the lineage of Adam and Eve.

## NEANDERTHAL EXTINCTION

If you search the Internet for when evolutionists think Neanderthals existed on Earth, you will find many theoretical timetables from 250,000 to less than 100,000 years ago. Some theories suggest Neanderthals existed at least 70,000 years ago and started to interbreed with humans about 47,000 years ago for about 7,000 years. Some think Neanderthals went extinct sometime after this. The reasons, however, are controversial and are still being researched.

Note: Because of the significant variation in the theories of evolution about Neanderthal existence, I am not providing websites. You can search the Internet if you want more details.

### Neanderthal Survival Compared to Modern Humans Theory

The most common ideas from evolutionists about the cause of the extinction of Neanderthals compared to their human co-inhabitants are as follows:

- Humans were better at acquiring and using resources to survive and thrive
- Larger human families and social groups enabled breeding for more offspring, so more humans lived and survived over time

- Higher birth and lower mortality rates and other demographic factors enhanced human survival
- Small, isolated groups of Neanderthals in colder climates decreased their survival rates

## Did They Really Become Extinct?

Some scientists think that Neanderthals did not become an extinct species. They believe, instead, that their comparatively small numbers were absorbed into the modern human gene pool. We know that people in many geographic groups worldwide have Neanderthal DNA. This fact could support their supposition.

## Scientific Conclusions about Neanderthals

We can conclude from science the following to be true about Neanderthals:

- They existed in the same period as human beings
- They were genetically close to humans to interbreed with them successfully
- They had many of the same behavioral and sentient characteristics as humans
- They were not evolutionary transitional species to modern humans

Definition: *Sentience* is the ability of humans to sense what is occurring within and around them. They are conscious of and can thoughtfully respond and interact with their environment. Self-awareness allows them to sense what is happening in their minds and hearts. Neanderthals were sentient beings.

## Biblical Conclusions about Neanderthals

From our assessment of the creation account in the Bible, we can theorize the following about Neanderthals:

- Neanderthals might have been created on Day Six as the same or a separate species of the human creation kind, or
- Neanderthals might have been the offspring of Noah's children after the flood
- The question remains: were they created in God's image?

Instructional Comment: Since Neanderthals interbred with humans, they were from the same creation kind. We saw earlier that hybridization experiments typically allow different species of the same kind to interbreed successfully. Successful hybridization usually does not occur with different species from different creation kinds. Therefore, we can conclude that Neanderthals and modern humans were of the same creation kind whether or not they were the offspring of Adam and Eve's or Noah's children. We can't know if they were made in God's image.

## CHAPTER QUESTIONS

DQ1: Why does Neanderthal DNA validate creation and disprove the theory that Neanderthals evolved into modern humans?

DQ2: In what ways do you think the discoveries of the Denisova Cave support the idea of creation and disprove evolution?

DQ3: Why do you think Neanderthals were created on Day Six but were not created in God's image?

DQ4: What do you think Noah and his family had 1 to 4 percent Neanderthal DNA they passed on to all humanity?

## DIG DEEP

DD1: Listen to the podcast by Trey Bowling with Dr. Brian Thomas, "The Soulless Hominid Theory: A Fatal Flaw in Old Earth Creationism." Explain the following: 1. Why did early drawings of Neanderthals affect people's ideas of them? 2. Why does Dr. Thomas say Neanderthals are human beings? Website: https://www.icr.org/article/15206.

## PERSONAL APPLICATION

PA1: If you had your DNA tested, the result would indicate you have between 1 and 4% Neanderthal DNA. Why would you accept or reject the fact that you are part Neanderthal?

# 19.

# Sixth Day: People (Nature and Image of God)

In this chapter, you will learn that only people were created in God's image on Day Six, when he made all land-based creatures. (We are uncertain if Neanderthals were created in God's image.) You will also learn that he created people with a three-fold nature: spirit, soul, and physical body.

> God said, "Let the land produce living creatures according to their kinds: cattle, creeping things, and wild animals, each according to its kind." It was so. God made the wild animals according to their kinds, the cattle according to their kinds, and all the creatures that creep along the ground according to their kinds. God saw that it was good. Then God said, "Let us make humankind in our image, after our likeness, so they may rule over the fish of the sea and the birds of the air, over the cattle, and over all the earth, and over all the creatures that move on the earth." God created humankind in his own image, in the image of God he created them, male and female he created them. (Genesis 1:24–27 NET)

Definition: *Humankind*: I use the words "people," "human beings," "humankind," "humanity," or "Homo sapiens" to represent both male and female genders. "Homo sapiens" is the scientific name for the human species. I don't use "mankind" except where various theories or scriptures use it.

## PEOPLE CREATED WITH A THREE-FOLD NATURE

The three-fold nature of people can be clearly seen in several New Testament scriptures:

> Now may the God of peace himself make you completely holy and may your *spirit* and *soul* and *body* be kept entirely blameless at the coming of our Lord Jesus Christ. (1 Thessalonians 5:23 NET, author's emphasis)

> For the word of God is living and active and sharper than any double-edged sword, piercing even to the point of dividing *soul from spirit*, and *joints from marrow*; it

is able to judge the desires and thoughts of the heart. (Hebrews 4:12 NET, author's emphasis)

Instructional Comment: In the above verse, you see the three-fold nature of people as a spirit, soul, and physical form. *Joints from marrow* represent the physical body. The fact that the human soul can be divided from the human spirit indicates that they are separate parts of human nature.

## IMAGE OF GOD IN HUMANKIND

Only God knows for sure what image he passed on to humankind. Perhaps we should approach this from a logical point of view rather than absolute biblical or scientific certainty.

The following is what I believe to be true about the image of God in human beings:

- It's not the physical nature of God they resemble
- It's a unique God-centered nature that represents (or reflects) his soul and spirit
- It comes from the soul and spirit of God's nature
- It includes God's unique gift of authority to rule as the Creator God's representative on Earth over all creation
- It's the spirit component that ultimately differentiates humans from all others as having the image of God (nothing else created has a spirit)
- His image does not include God's sovereign, supreme, divine attributes
- Angels and animals who have a soul do not have the image of God

### People Share in the Divine Nature of God

The following verse is troubling when considering the image of God.

> And because of His glory and excellence, He has given us great and precious promises. These are the promises that *enable you to share His divine nature* and escape the world's corruption caused by human desires. (2 Peter 1:4 NLT, author's emphasis)

Instructional Comment: Since people are created beings and not the Creator, they cannot be divine in their essential nature. They only resemble their Creator's divine nature.

---

One way people "share his divine nature" is by sharing God's morality. In other words, people resemble his moral nature.

---

## REASONS PEOPLE WERE CREATED IN GOD'S IMAGE

People are unique among all the creations designed by our Intelligent Designer God. According to the Bible, Jesus (following the Father's design) created humankind in the image of God. In so doing, they were uniquely given a God-centered nature.

The Creator God did this so people with a God-centered nature would:

- Uniquely bear the image of God
- Become his children
- Live eternally with God
- Represent God in all his creation
- Rule over God's creation
- Become the human vehicle of spiritual life on Earth

## PEOPLE ARE THE REPRESENTATION OF GOD ON EARTH

Carmen Joy Imes wrote *Being God's Image, Why Creation Still Matters*. She says because the Creator God made people in his image, they are his representatives on Earth. They are the physical means on Earth by which he exhibits his presence, power, and authority over his creation. As his image bearers, people reflect something of God's character and nature to the rest of creation.

> *God created man in His own image*, in the image of God He created him; male and female He created them. God blessed them; and God said to them, "Be fruitful and multiply, and *fill the earth, and subdue it; and rule* over the fish of the sea and over the birds of the sky and over every living thing that moves on the earth." (Genesis 1:27–28 NASB, author's emphasis)

## JESUS CHRIST INCARNATE IS THE FULL IMAGE OF GOD

When Jesus Christ came to Earth as the Son of God and the Son of Man, he was the fullness of God and the fullness of man. He did not just resemble the image of God. In other words, he was the full image of God as well as the full image of man as the Father's visible presence on Earth.

We see this fullness of the image of God in Jesus in the following verses:

> Christ is the visible image of the invisible God. (Colossians 1:15 NLT)

> For God in all His fullness was pleased to live in Christ. (Colossians 1:19 NLT)

# HUMAN SPIRIT: UNIQUE TO PEOPLE

*In addition to the Creator God's authority, I believe it's the human spirit that makes people unique from all other creatures.* The spirit is the essential part of the three-fold nature of people that enables them to have the image of God. Without a spirit, people would be like animals and would not have the image of God.

Instructional Comment: If Neanderthals were made in the image of God, they would have had a spirit like modern humans.

## Purpose of the Spirit in People

I believe Scripture tells us the purpose of the human spirit given to humankind was to enable them to commune and communicate with their Creator God. However, they lost this ability after the rebellion of Adam and Eve. These spiritual abilities are only restored when people accept Jesus Christ as their Savior and Lord. At that moment, the Holy Spirit comes to live with them. They are born-again. They have the restored ability to commune and communicate with God through his indwelling Spirit. He causes a person to become a spiritual being in union with the spiritual God of the Bible.

Following are some scriptures that demonstrate the Holy Spirit speaks to the born-again human spirit to reveal the things of God:

> That is what the Scriptures mean when they say, "No eye has seen, no ear has heard, and no mind has imagined what God has prepared for those who love Him." But it was to us that God revealed these things by His Spirit. For His Spirit searches out everything and shows us God's deep secrets. For who among men knows the things of a man except the man's spirit within him? So too, no one knows the things of God except the Spirit of God. Now we have not received the spirit of the world, but the Spirit who is from God, so that we may know the things that are freely given to us by God. And we speak about these things, not with words taught us by human wisdom, but with those taught by the Spirit, explaining spiritual things to spiritual people. (1 Corinthians 2:9–13 NLT)

## Non-Believers' Spirit Is Dead to God

Ephesians 2:5–6 says that non-believers are dead to God due to their unforgiven sins. But what was dead in them before they were saved? Not their physical body nor their soul. They were physically alive, and they could think and feel, so it was their spirit that was dead to God. In what way was their spirit dead? If something is dead, it cannot function. It needs resurrection to become alive and functional. I believe the Holy Spirit makes the human spirit alive to God in born-again believers. They then become spiritual beings. It seems Adam and Eve's rebellion resulted in the death of the human spirit to God.

Even though we were dead because of our sins, He gave us life when He raised Christ from the dead. (It is only by God's grace that you have been saved!) For He raised us from the dead along with Christ and seated us with Him in the heavenly realms because we are united with Christ Jesus. (Ephesians 2:5–6 NLT)

## HUMAN SOUL

The human soul reflects who a person is. A physical body may be pleasing to sight, but it is not the person. It's the exterior shell in which the soul and spirit reside. The real person is on the inside, not the outside. We talk about the "personality" of a person. It reflects the individual's uniqueness, which is developed over time based on the functions of the soul, experiences in life, teaching, impacts of culture, and so on.

Some simplify the human soul as consisting of the mind, emotions, and will. I break this down a little further as follows:

- Emotions (feelings)
- Will
- Reasoning/Thinking
- Memories
- Conscience (morality)

As you can see, these are all non-physical characteristics of humankind.

Instructional Comment: Many characteristics of the human soul can be seen in animals, especially in primates such as apes and chimpanzees. However, there is no scientific evidence that animals have a conscience, allowing them to discern right from wrong or be self-aware. Animals also do not have a spirit that enables them to commune and communicate with God (as born-again Christians can do).

## SOUL AND BODY ARE INTRICATELY DESIGNED

When Adam was made from the dust of the Earth, he (and humankind) was given an intricately designed body and soul. We see this intricate design at human conception in the following verses:

> For You formed my inward parts; You wove me in my mother's womb. I will give thanks to You, for I am fearfully and wonderfully made; Wonderful are Your works, And my soul knows it very well. My frame was not hidden from You, When I was made in secret, And skillfully wrought in the depths of the earth. (Psalm 139:13–15 NASB)

> The Hebrew word for *inward parts* is *kilyâh*; which refers to the soul, the inner person, being formed in the womb.

The Hebrew word for *frame* is *'ôtsem*; which refers to the physical body being formed in the womb.

## HUMAN PHYSICAL BODY

The human body (and that of all animals) is the outer shell used by the brain and mind to interact with the surrounding physical world. People interact with it through the five God-given physical senses: sight, hearing, touch, smell, and taste. The nerves in the related parts of the body relay signals to the brain, which interprets them and informs the mind on how to respond.

### Purpose of Human Body

As you saw in the chapter, "God's Purpose for Creation," that God gave humankind a physical body to enable people to live and work in the physical world created for them.

We see this in the following verses:

> For the LORD is God, and He created the heavens and earth and put everything in place. He made the world to be lived in, not to be a place of empty chaos. (Isaiah 45:18 NLT)

> The LORD God took the man and put him in the garden of Eden to work it and keep it. (Genesis 2:15 ESV)

We see in the following verses that God gave humankind a physical body to be able to exercise his authority over creation:

> When I look at the night sky and see the work of Your fingers—the moon and the stars You set in place—what are mere mortals that You should think about them, human beings that You should care for them? . . . You gave them charge of everything You made, putting all things under their authority. (Psalm 8:3–6 NLT)

### Natural Elements of the Human Body

The Father designed, and Jesus created the human body from the dust of the ground. The body is, therefore, composed of the natural elements Jesus placed in the ground when he created the Earth out of nothing.

Following is a look at the natural elements of the Periodic Table that comprise the human body. Only 21 elements of the 92 known natural elements comprise the human body. Four elements comprise about 96 percent of the body, while oxygen is 65 percent of it.

- Oxygen
- Carbon

- Hydrogen
- Nitrogen

The following additional seven natural elements (essential minerals) compose about 3 percent of the body:

- Calcium
- Phosphorus
- Potassium
- Sulfur
- Sodium
- Chlorine
- Magnesium

Note: The human body also consists of many trace elements. Some of these are essential for the body to function, while others are simply beneficial. Still others have no known function, while some are toxic in excess.

### Life in This World Is Impossible without a Physical Body

Isn't it unimaginable to think that our Creator God would have made people without a physical body? How would people be able to interact with the physical world they were made to live in without it? How would we eat, drink, reproduce, and work without this amazing provision from the all-knowing Creator God?

## CHAPTER QUESTIONS

DQ1: If people were not created in the image of God, how would they be different from all other creatures?

DQ2: Why did Jesus Christ incarnate exist on Earth in the full image of God?

DQ3: What part of the three-fold nature of people is required for people to be made in the image of God?

DQ4: Why is life on Earth for people impossible without a physical body?

## DIG DEEP

DD1: Why were our bodies created from the soil of the Earth (natural elements) and not out of nothing (*Ex Nihilo*)?

## PERSONAL APPLICATION

PA1: If you are a born-again follower of Jesus, describe how you sense the presence of the Holy Spirit and experience his communications.

# 20.

# God's Nature

The God of the Bible is the Triune God, consisting of the Father, the Son, Jesus Christ, and the Holy Spirit. Each is equally divine and fully God in their essential being and nature. The image of God given to humankind comes from the nature of all three persons.

Note: This chapter contains my opinions about the nature of God. These opinions are based on how I understand Bible verses related to these topics. Your opinions may differ from mine.

## GOD'S NATURE INCLUDES THREE UNIQUE DIVINE ATTRIBUTES

God's image (resemblance) is what was passed on to Adam and Eve at their creation. Adam and Eve were created to be perfect, eternal beings representing the Trinity of God on Earth. They were not created to be God or equal to the Triune God. Therefore, the image of God in people is not the Trinity's divine nature.

### Three Unique Divine Attributes

The following three unique attributes are part of God's divine nature but not part of his image given to people. Only the three persons of God can know everything, be everywhere, and exert unimaginable supernatural power everywhere.

### Omniscient (Sees, Knows Everything)

Because the Triune God lives in eternity (in the past, present, and future, all at once), there is no boundary of space and time for them. They see and know everything.

> You know when I sit down or stand up. You know my thoughts even when I'm far away. You see me when I travel and when I rest at home. You know everything I do. You know what I am going to say even before I say it, LORD. (Psalm 139:2–4 NLT)

### Omnipresent (Present Everywhere)

The Triune God is everywhere. This supernatural presence includes the universe and Earth.

> "Am I a God who is only close at hand?" says the LORD. "No, I am far away at the same time. Can anyone hide from Me in a secret place? Am I not everywhere in all the heavens and earth?" says the LORD. (Jeremiah 23:23–24 NLT)

### Omnipotent (All-Powerful)

The Triune God is sovereign in power. Nothing or no one can prevent their will from being done. Their power is infinite and limitless.

> All the people of the earth are nothing compared to Him. He does as He pleases among the angels of heaven and among the people of the earth. No one can stop Him or say to Him, "What do you mean by doing these things?" (Daniel 4:35 NLT)

## DOES GOD'S NATURE INCLUDE A SOUL, SPIRIT, AND BODY?

In the previous chapter, I described people created in God's image as having a three-fold nature consisting of a soul, a spirit, and a physical body.

> So God created human beings in His own image. In the image of God He created them; male and female He created them. (Genesis 1:27 NLT)

> The Hebrew word for *image* is *tselem*; which means a *resemblance* or representative figure, such as an *image*.

> Now may the God of peace himself make you completely holy and may your *spirit* and *soul* and *body* be kept entirely blameless at the coming of our Lord Jesus Christ. (1 Thessalonians 5:23 NET, author's emphasis)

A reasonable question is, "Does each person of God have a three-fold nature consisting of a soul, a spirit, and a physical form? If so, is the divine three-fold nature the same for each person of God?

## THREE-FOLD NATURE OF GOD

Let's look at Scripture to see if we can determine if each person of God has a three-fold nature consisting of a soul, a spirit, and a physical form. I think you will agree they each have a soul and spirit as part of their divine nature. The rest of this chapter focuses on differences in their physical nature.

> The Creator God gave people a physical body as part of their three-fold nature to exist within the physical world. Since the Triune God does not exist within the physical world, he does not need a physical body.

Following is a summary of how I understand Scripture about a divine three-fold nature for each person of God. Only a physical form or material substance is different for each of them.

- Soul: Each person of God has a soul
- Spirit: Each person of God has a spirit
- Physical form or material substance: This is different for each person of God

## Soul

I think most people would agree that each person of God has a soul. Throughout the Bible, we see each person of God exhibiting characteristics of a soul. These are similar characteristics we see in people.

As I stated in the previous chapter, "Sixth Day: People (Nature and Image of God)," I believe the human soul was created to be like the divine soul, consisting of the following:

- Emotions (feelings)
- Will
- Reasoning/Thinking
- Memories
- Conscience (morality)

## Spirit

*In addition to the Creator God's authority, I believe it's the human spirit that makes people unique from all other creatures.* Each person of God must have a spirit and be a spiritual being.

The Holy Spirit, the Spirit of God, is a spirit that has no physical form or material substance. Because of this, he is able to live within every person who accepts Jesus as Savior and Lord.

> But when the Helper comes, whom I will send to you from the Father, the Spirit *(pheuma)* of truth, who proceeds from the Father, he will bear witness about me. (John 15:26 ESV, author's emphasis)

> The Greek word for *spirit*, *pneuma*, is used to describe God the Father and the Holy Spirit. It refers to a current of air, breath, or breeze.

The Father God is a spirit in his divine nature (his essential being). But is he the same invisible being as the Holy Spirit?

> For God *(Father)* is Spirit *(pneuma)*, so those who worship Him must worship in spirit *(pneuma)* and in truth. (John 4:24 NLT, author's emphasis)

Does Jesus Christ also have a spirit as part of his divine nature? Yes, he has since he is fully and equally God, as we see in the following verse:

> For in him the whole fullness of deity dwells bodily. (Colossians 2:9 ESV)

Since Jesus is entirely divine and God in every way, he must be spirit in his divine nature, even though he exists in Heaven in an eternal, visible, physical form.

Instructional Comment: We know the incarnate Jesus Christ rose from the dead and ascended to Heaven with a supernatural, eternal physical body. And he now sits on his throne in Heaven next to the Father God.

## CHAPTER QUESTIONS

DQ1: Why does the image of God given to humankind come from the nature of all three persons, not just the Father God?

DQ2: What are the three divine attributes only the Triune God has?

DQ3: Describe the physical nature of each person of God.

DQ4: We know the Father God and Holy Spirit are spirit in their nature. Why must Jesus also be spirit in his nature?

## DIG DEEP

DD1: Why is the nature of the Trinity of God a mystery? Why can't people clearly and fully understand the nature of the Trinity?

## PERSONAL APPLICATION

PA1: How does knowing the Triune God's three divine attributes affect your prayers?

# 21.

## Manifestations of God

Let's look at how the Trinity demonstrates their existence and involvement with people.

### THEOPHANY AND CHRISTOPHANY: PHYSICAL MANIFESTATIONS OF GOD

The Holy Spirit is God with an invisible presence on Earth (except for a few physical manifestations). Since the Father God and Jesus do not exist on Earth in a physical manner (except for the incarnation of Jesus), they choose to interact with people from their dwelling in Heaven.

In both the Old and New Testaments, we see that each person of God manifests himself on Earth in physical ways. For example, in the New Testament, we see the manifestation of the Father as he split apart the heavens and spoke to people at the baptism of Jesus (Mark 1:9–11). We see the Holy Spirit manifesting himself in the physical form of a dove at this time. And we see Jesus manifesting himself from Heaven when he created a bright light and spoke to Saul (Acts 9:1–5).

> Each person of God has manifested himself on Earth in physical ways. These physical manifestations of the Father God and Jesus are called "theophanies," while a manifestation of Jesus is a "Christophany."

These physical manifestations of the Father God and Jesus are called "theophanies," while a manifestation of Jesus is a "Christophany." The manifestation of the Holy Spirit in a physical manner is not typically seen as a theophany. However, it does fit the general definition of a theophany.

Definition: *Theophany* is a temporary appearance of the Father God or Jesus on Earth involving a visual manifestation to capture people's attention. Since the Holy Spirit is on

Earth (he is in Heaven also), his physical manifestations, such as a dove, are not considered theophanies.

*Christophany* is a temporary appearance of Jesus Christ on Earth with a physical body or another physical manifestation. Sometimes, these are referred to as "pre-incarnate" appearances of Jesus in the Old Testament.

*Incarnation of Jesus Christ* is the temporary appearance of Jesus Christ on Earth with a physical body as described in the New Testament. When Jesus ascended to Heaven, he went with his resurrected incarnate, eternal, and permanent physical body.

### My Theophany of the Father

In my previous book *Who Is This God? A Handbook for Life with Him*, I describe my only encounter with a manifestation of the Father God in a theophany. I experienced his voice, presence, and muted glorious light in my bedroom as he communicated his will in a particular matter. Knowing that our limited, fragile human nature cannot see him and live, he revealed himself dimly in all these expressions. Yet, it was very clear that it was the Father and not Jesus nor the Holy Spirit I was encountering.

### Different Types of Theophanies and Christopnaies

The Father God and Jesus appear to, and communicate with, people in the Bible in many different ways. For example, they do so in thunderstorms, clouds, smoke, pillars of fire, chariots, and more. Jesus appears physically on several occasions in the Old Testament when he is referred to as "the angel of the Lord" (Genesis 16:7, 22:11). Some scholars believe many Old Testament theophanies are Christophanies, manifestations of Jesus and not the Father. One clear exception is when the Father God tells Moses that he cannot see him and live (Exodus 33:18–23). Yet, it seems that in Exodus 33:7–11, it is the pre-incarnate Jesus (Christophany) who speaks with Moses face-to-face. Because it is Jesus Moses is seeing, not the Father, Moses lives.

> The LORD (*Jesus*) would speak to Moses face to face, the way a person speaks to a friend. (Exodus 33:11 NET, author's emphasis)

Let's focus on the Father God to consider whether we think he has a visible, physical form in Heaven. This controversial idea challenges our understanding of his divine nature.

Note: The following contains my opinions about the nature of God. These opinions are based on how I understand Bible verses related to these topics. Your opinions may differ from mine.

## BIBLICAL FIGURATIVE LANGUAGE

The Bible has many scriptures that are figures of speech. These are expressions not meant to be understood in a literal manner. Instead, they are an expression meant to describe an aspect of biblical truth. One biblical scholar identified 217 different types of figures of speech in the Bible. Twelve common biblical figures of speech are: Anthropomorphism, metaphor, personification, symbolism, simile, hyperbole, allegory, parable, irony, euphemism, oxymoron, and synecdoche.

We have seen that each person of God can manifest himself in physical appearances meant to be understood literally. What else does the Bible tell us about how they express their physical nature to people?

Definition: *Figurative language* (figures of speech) in the Bible represents an idea that is not literal.

### Figures of Speech and the Father of God?

We know the Father manifests himself on Earth through theophanies. We also know he never leaves Heaven to come to Earth until the new Earth (Revelation chapters 21 and 22). But what about the following Bible verses (and others) that ascribe physical attributes to him? Are these figurative language not meant to be understood in a literal manner?

> The LORD your God—His greatness, His mighty hand and His outstretched arm. (Deuteronomy 11:2 NASB)
>
> Behold, the eyes of the Lord GOD are on the sinful kingdom. (Amos 9:8 NASB)

Many Christians see these as figures of speech that indicate an action the Father God does from Heaven and not literal statements about him having physical hands, arms, or eyes.

## ANTHROPOMORPHISM EXPLAINS SOME FIGURATIVE LANGUAGE

Some refer to verses like the above as "anthropomorphisms." This term, in particular, is used to describe the Father God as having features of the human body and soul so people can better understand and relate to him. They see these verses as figurative language that does not imply the Father God exists in a visible, physical form in Heaven. They believe these verses are in the Bible to help people more easily relate to him as a personal Father God who is a spirit.

## DOES THE FATHER GOD EXIST IN A VISIBLE, PHYSICAL FORM IN HEAVEN?

Biblical scholars have debated this question for decades. Christians have varied beliefs about this. So, I can only provide my opinions based on my understanding of Scripture.

Instructional Comment: Let me begin by saying that I do not believe the Father God is a physical being in Heaven with a physical body, not even one like Jesus' eternal physical body. However, I suspect he is not an invisible wind or vapor in Heaven. I wonder if the following verses and discussion might indicate he has some material form in Heaven (but not one like Jesus or angels)?

Important: We know from the Bible that the Father God has never left Heaven to come to Earth. Instead, he sends Jesus or angels or uses people to do his will on Earth.

### Jesus Saw the Father in Heaven

In the following verse, Jesus tells us he saw the Father in Heaven before he came to Earth at his incarnation. Does this imply the Father has a visible form that he saw in Heaven?

> Not that anyone has ever seen the Father; only I *(Jesus)*, who was sent from God, have seen Him. (John 6:46 NET, author's emphasis)

### Cannot See Father God's Face and Live

If the Father God was an amorphous, invisible being in Heaven, why would he tell Moses he could not see him and live? If he is invisible to people, he cannot be seen. Or is this also figurative language?

> Then Moses said, "I pray You, show me Your glory!" And He said, "I Myself will make all My goodness pass before you, and will proclaim the name of the LORD before you; and I will be gracious to whom I will be gracious, and will show compassion on whom I will show compassion." But He said, *"You cannot see My face, for no man can see Me and live!"* Then the LORD said, "Behold, there is a place by Me, and you shall stand there on the rock; and it will come about, while My glory is passing by, that I will put you in the cleft of the rock and cover you with My hand until I have passed by. Then I will take My hand away and you shall see My back, but My face shall not be seen." (Exodus 33:18–23 NASB, author's emphasis)

Instructional Comment: I suspect this is not figurative language. Notice the Father's response to his beloved servant, Moses. He grants Moses' desire to have a visual encounter with him, but only in a minute brief manner referred to as his "goodness" that quickly passes by Moses.

### Corrupted Nature of Humankind

People in our fallen human nature cannot see the glory or presence of the Father God and live. Our physical, mental, and spiritual natures are too weak and intolerant. This is why we would die if he showed us his presence on Earth. Only in our resurrected spirit, soul, and body in Heaven and the new Earth will we see him "face to face."

### Father God Sits on His Throne in Heaven

How can the Father God sit on his throne in Heaven if he is like the wind with no physical substance?

> Jesus . . . sat down at the right hand of the throne of God. (Hebrews 12:2 NASB)

> They are before the throne of God and serve him day and night in his temple; and he who sits on the throne will shelter them with his presence. (Revelation 7:15 NASB)

## FATHER GOD ON THE NEW EARTH

In Revelation chapter 21, we are told that the Father God and Jesus will have visible, physical forms on the new Earth because believers there will see their faces and interact with them.

> Mystery: If the Father God does not currently exist in a visible, physical form in Heaven, wouldn't he need to change his divine nature to become a visible, physical being when he comes to the new Earth to live with believers?

> There will no longer be any curse; and the throne of God and of the Lamb will be in it, and *His bond-servants will serve Him; they will see His face*, and His name will be on their foreheads. And there will no longer be any night; and they will not have need of the light of a lamp nor the light of the sun, because the Lord God will illumine them; and they will reign forever and ever. (Revelation 22:3–5 NASB, author's emphasis)

> Then I saw a new heaven and a new earth, for the first heaven and earth had ceased to exist, and the sea existed no more. And I saw the holy city—the new Jerusalem—descending out of heaven from God, made ready like a bride adorned for her husband. And I heard a loud voice from the throne saying: "Look! *The residence of God is among human beings. He will live among them, and they will be his people, and God himself will be with them.* (Revelation 21:1–3 NET, author's emphasis)

These verses in Revelation, chapters 21 and 22, seem to describe the Father's and Jesus' visible, physical presence on the new Earth as literal, not figurative, or as anthropomorphisms.

## ANGELS ARE SPIRITS

The Father God's angels are a spirit (*pneuma*). They are supernatural, spiritual beings created from spiritual elements (not from the natural elements of the Earth). Yet, they seem to manifest a physical form when they stand before the Father God in Heaven. They can also be seen on Earth when the Father wants it. There are many examples in the Old and New Testaments about angels appearing in a physical form to people. Shouldn't we consider these to be literal experiences?

## Angel Gabriel

The following verse tells us that Gabriel was an individual angel who stands in the presence of the Father God in Heaven. We also learn from it that he was sent to Earth as a visible angel to be seen and heard on several occasions.

> An angel of the Lord, standing on the right side of the altar of incense, appeared to him. And Zechariah, visibly shaken when he saw the angel, was seized with fear. But the angel said to him, "Do not be afraid, Zechariah, for your prayer has been heard, and your wife Elizabeth will bear you a son; you will name him John." . . . The angel answered him, "I am Gabriel, who stands in the presence of God, and I was sent to speak to you and to bring you this good news. (Luke 1:11–19 NET)

Important: It is apparent from Scripture that every supernatural being in the Bible can be manifested in a physical form on Earth. This seems to include the Father God.

## CHAPTER QUESTIONS

DQ1: Describe the difference between a theophany and a christophany.

DQ2: How does biblical figurative language differ from biblical literal language?

DQ3: Why do you believe the Father does or does not exist in Heaven with some material substance?

DQ4: How are angels different in spirit than the Holy Spirit and the Father God?

## DIG DEEP

DD1: From the scriptures, explain why the Father God may have some form of a physical being in Heaven.

## PERSONAL APPLICATION

PA1: How does your personal relationship with the Father God change if you believe he is a formless entity (like the wind)?

# 22.

# God's Image Restored after the Fall

## CORRUPTED IMAGE BEARERS

People were created in God's image and originally had a God-centered nature. However, they lost that God-centered nature with the fall of Adam and Eve. It was replaced with a corrupted self-centered nature ("old self" or "old nature"). This fallen nature has power over their minds and lives instead of the Creator God.

> Put off your old self, which belongs to your former manner of life and is corrupt through deceitful desires (Ephesians 4:22 ESV)

The Apostle Paul in the New Testament describes the fallen old nature of people when they are born into this world. They do not have the original unique spiritual connection to God. He describes this fallen self-centered nature in the following verses:

> So this I say, and affirm together with the Lord, that you walk no longer just as the Gentiles also walk, in the futility of their mind, being darkened in their understanding, excluded from the life of God because of the ignorance that is in them, because of the hardness of their heart. (Ephesians 4:17–18 NASB)

## RESTORED GOD-CENTERED NATURE AND IMAGE

When people accept Jesus Christ as their Savior and Lord and are born-again, they are adopted into our Heavenly Father's family as his children. When this occurs, they are a new creation, created to be like Jesus Christ! The image of God is restored in them. Their fallen, corrupted old nature no longer has complete power over their minds, hearts, and lives. They once again have a God-centered nature.

> When people accept Jesus Christ as their Savior and Lord, they are born-again and become a new creation, created to be like Jesus Christ! The image of God is restored in them.

> Therefore if anyone is in Christ, he is a new creature; the old things passed away; behold, new things have come. (2 Corinthians 5:17 NASB)

> The Greek for *new* is *kainos*; new kind, unprecedented, novel, unheard of.

> Put on the new self, which in the likeness of God has been created in righteousness and holiness of the truth. (Ephesians 4:24 NASB)

Note: Born-again followers of Jesus still have their self-centered old nature. But they now have a new nature to help them live for Christ.

Followers of Christ are being renewed in their minds and transformed in their hearts as they regain the image of God lost with the fall of Adam and Eve.

> Have put on the new self who is being renewed to a true knowledge according to the image of the One who created him. (Colossians 3:10 NASB)

Believers must choose to live by the indwelling Holy Spirit and learn to live by their newly created nature. Learning to live by this new nature is not easy. It is a difficult process and sometimes a battle over time. But Christians have someone to help and enable them. It's the indwelling Holy Spirit (John 14:26).

> For those who live according to the flesh *(old nature)* set their minds on the things of the flesh, but those who live according to the Spirit *(new nature)* set their minds on the things of the Spirit. For to set the mind on the flesh is death, but to set the mind on the Spirit is life and peace. For the mind that is set on the flesh is hostile to God, for it does not submit to God's law; indeed, it cannot. Those who are in the flesh cannot please God. (Romans 8:5–8 ESV, author's emphasis)

The Holy Spirit transforms the believer's old nature into Christ's new nature as they strive to listen to him and obey him. This new nature produces a lifestyle and behaviors that God values and rewards. This new nature is visible through the "Fruit of the Spirit."

> But the Holy Spirit produces this kind of fruit in our lives: love, joy, peace, patience, kindness, goodness, faithfulness, gentleness, and self-control. (Galatians 5:22–23 NLT)

## CREATION MANDATE AFTER THE FALL

A mandate is an order or commission to do something with the authority of the one giving the mandate. The "creation mandate" was given by God to Adam and Eve and their offspring to rule over and care for his creation. This is what we see in the next verse:

God blessed them and said to them, "Be fruitful and increase in number; fill the earth and subdue it. Rule over the fish in the sea and the birds in the sky and over every living creature that moves on the ground." (Genesis 1:28 NIV)

*Rule*, Hebrew is *radah*; to have dominion, rule, dominate.

Albert Barnes[1] says the following about the creation mandate:

> Hence, it is necessary that he should receive from high heaven a formal charter of right over the things that were made for man. He is therefore authorized, by the word of the Creator, to exercise his power in subduing the earth and ruling over the animal kingdom. This is the meet sequel of his being created in the image of God. Being formed for dominion, the earth and its various products and inhabitants are assigned to him for the display of his powers. The subduing and ruling refer not to the mere supply of his natural needs, for which provision is made in the following verse, but to the accomplishment of his various purposes of science and beneficence. (Albert Barnes' Notes on the Bible)

## Does the Creation Mandate Exist Today?

Do Christians today still have the creation mandate to rule over and care for God's creation, even though creation was corrupted by the sin of Adam and Eve? Yes, I believe they do.

In the following verse, we see that Jesus has authority over everything:

> Then Jesus came up and said to them, "All authority in heaven and on earth has been given to me." (Matthew 28:18 NET)

Jesus said he has given his followers authority to accomplish the Father's will on Earth. I believe this applies to carrying out the creation mandate. The restored image of God includes his authority to rule and care for creation.

> Look, I have given you authority to tread on snakes and scorpions and on the full force of the enemy, and nothing will hurt you. (Luke 10:19 NET)

However, non-believers have not regained God's authority to rule and care for his creation.

> Then to Adam He said, . . . "Cursed is the ground because of you; In toil you will eat of it All the days of your life." (Genesis 3:17 NASB)

> For the sinful nature is always hostile to God. It never did obey God's laws, and it never will. That's why those who are still under the control of their sinful nature can never please God. (Romans 8:7–8 NLT)

It is clear from Scripture that followers of Jesus Christ have been given the authority of Jesus Christ to exercise his will on Earth. We see this requirement in the creation mandate.

---

1. Barnes, *Notes*, Genesis 1:28.

## CHAPTER QUESTIONS

DQ1: Why is "putting off the old self" involved in having God's image restored?

DQ2: When is the image of God's nature restored to a person?

DQ3: Why does the restored image of God require a uniquely new nature (*kainos*) in believers?

DQ4: Why do you think the creation mandate still exists for Christians?

## DIG DEEP

DD1: What is the new nature of God in born-again believers? What are some characteristics that people can see?

## PERSONAL APPLICATION

PA1: In what ways would a Christian exercise Jesus' biblical authority to rule over and care for God's creation?

# 23.

## Seventh Day: Sabbath

This chapter answers the questions, "What is the Sabbath, why was it created, and who is required to obey it?"

> On the seventh day God *(Elohim)* had finished His work of creation, so He rested from all His work. And God blessed the seventh day and declared it holy, because it was the day when He rested from all His work of creation. (Genesis 2:2–3 NLT, author's emphasis)

The Holy Spirit tells us the following about the Sabbath rest:

- The Creator God rested on the seventh day of the week
- The work of creation was completed
- The seventh day was blessed and declared holy (set aside by God, for God)

*Holy*, Hebrew is *qâdash*; consecrate, dedicate, sanctify.

In the above verses in Genesis, God does not tell people to rest on the seventh day of the week. Instead, he states why he stopped his first creation: He was finished with it. (There is also his creation of the new heavens and Earth that occur at the end of time.)

Important: The God of the Bible is El Shaddai and God Almighty. He is all-powerful, all-knowing, and ever-present. He never gets tired and never loses interest in what he is doing.

### JEWISH PEOPLE AND THE SABBATH

The Holy Spirit spoke through Moses to write the first five books of the Old Testament, referred to as the "Torah" and "Pentateuch." In these books, God defined how the people of Israel were to live and obey him. This included 613 commandments concerning civil, ceremonial, and moral laws. These are called the Law of Moses (Mosaic Laws) and are the basis for the Old Covenant with Israel.

Important: God made the Old Covenant only with the people of Israel. All 613 commandments apply only to Israel. They do not apply to born-again followers of Jesus Christ since Jesus fulfilled the Law of Moses.

## TEN COMMANDMENTS

The Ten Commandments are included in the 613 commandments. They are a shortlist of concise statements about how the people of Israel were to interact with God and other people. The Fourth Commandment provides the reason why God required the people of Israel to rest on the Sabbath day.

Following are the Ten Commandments from the Old Testament of the Bible, quoted from Deuteronomy 5:6–22:

> I am the LORD your God, who rescued you from the land of Egypt, the place of your slavery.
>
> Commandment One: You must not have any other god but Me.
>
> Commandment Two: You must not make for yourself an idol of any kind, or an image of anything in the heavens or on the Earth or in the sea. You must not bow down to them or worship them, for I, the LORD your God, am a jealous God who will not tolerate your affection for any other gods. I lay the sins of the parents upon their children; the entire family is affected—even children in the third and fourth generations of those who reject Me. But I lavish unfailing love for a thousand generations on those who love Me and obey My commands.
>
> Commandment Three: You must not misuse the name of the LORD your God. The LORD will not let you go unpunished if you misuse his name.
>
> *Commandment Four: Observe the Sabbath day by keeping it holy, as the LORD your God has commanded you.* You have six days each week for your ordinary work, but the seventh day is a Sabbath day of rest dedicated to the LORD your God. On that day no one in your household may do any work. This includes you, your sons and daughters, your male and female servants, your oxen and donkeys and other livestock, and any foreigners living among you. All your male and female servants must rest as you do. *Remember that you were once slaves in Egypt, but the LORD your God brought you out with His strong hand and powerful arm. That is why the LORD your God has commanded you to rest on the Sabbath day.*
>
> Commandment Five: Honor your father and mother, as the LORD your God commanded you. Then you will live a long, full life in the land the LORD your God is giving you.
>
> Commandment Six: You must not murder.
>
> Commandment Seven: You must not commit adultery.
>
> Commandment Eight: You must not steal.
>
> Commandment Nine: You must not testify falsely against your neighbor.

Commandment Ten: You must not covet your neighbor's wife. You must not covet your neighbor's house or land, male or female servant, ox or donkey, or anything else that belongs to your neighbor.

The LORD spoke these words to all of you assembled there at the foot of the mountain. He spoke with a loud voice from the heart of the fire, surrounded by clouds and deep darkness. This was all He said at that time, and He wrote His words on two stone tablets and gave them to me. (Deuteronomy 5:6–22 NLT, author's emphasis)

## FOURTH COMMANDMENT

In the Fourth Commandment, God identifies who is required to honor the Sabbath and why they are to do so:

Remember that you were once slaves in Egypt, but the LORD your God brought you out with His strong hand and powerful arm. That is why the LORD your God has commanded you to rest on the Sabbath day (Deuteronomy 5:12–15 NLT).

He reiterates the who and why for keeping the Sabbath in several other scriptures in the Old Testament.

Tell the people of Israel: "Be careful to keep My Sabbath day, for the Sabbath is a sign of the covenant between Me and you from generation to generation. It is given so you may know that I am the LORD, who makes you holy." (Exodus 31:13 NLT)

Remember that you were once slaves in Egypt, but the LORD your God brought you out with His strong hand and powerful arm. That is why the LORD your God has commanded you to rest on the Sabbath day. (Deuteronomy 5:15 NLT)

Note: The Jewish Sabbath is Saturday, the last day of the week.

## OLD COVENANT AND THE SABBATH

The commandment for the people of Israel to keep the Sabbath was part of the Old Covenant. The Old Testament required complete obedience to every law of God, including the Sabbath. Its purpose was to show that people could not perfectly obey all the Laws to earn eternal life with God (Galatians 3:23–26). Instead, the Old Covenant pointed to the need for the New Covenant based on grace and faith, not people's efforts. The purpose of the New Covenant is to offer all people salvation as the free gift of God's grace (unmerited favor) and to set them free from religious efforts to please God and earn his favor.

For by grace you have been saved through faith. And this is not your own doing; it is the gift of God, a result of works, so that no one may boast. (Ephesians 2:8–9 ESV)

Definition: *"Old Testament"* and *"Old Covenant"* are not the same. The Old Testament includes 39 books from Genesis through Malachi. The Old Covenant is God's contract with his people, the Israelites only. It is based on the Law of Moses found in the Torah, the first five books of the Old Testament.

*New Covenant* was initiated by Jesus at the "Last Supper" to replace (not erase) the need for the Old Covenant. This fulfillment was for all people, Jews and Gentiles everywhere.

## JESUS FULFILLED THE OLD COVENANT

Jesus Christ is the only one who died and rose again to fulfill the legal requirements of the Old Covenant, including keeping the Sabbath. He did this to provide the free gift of salvation and eternal life to everyone who would accept him as Savior and Lord. This Good News is the New Covenant Jesus instituted at the "Last Supper" (Luke 22:14–20). Therefore, he is the only Mediator for this New Covenant (agreement or contract) between his followers and his Father God.

---

Worldview: With his death and resurrection, Jesus Christ fulfilled all the Old Testament commandments, including keeping the Sabbath. No one is required to keep the sabbath.

---

When Jesus came to this Earth, died, and rose again, he fulfilled all the Old Testament commandments, including the Ten Commandments.

> Don't misunderstand why I have come. I did not come to abolish the law of Moses or the writings of the prophets. No, I came to accomplish their purpose. (Matthew 5:17 NLT)

> Greek for *accomplish (or fulfill)* is *plēroō*; complete, end, expire, fill (up), fulfill, (be, make) full (come), perfect

> Then He said, "Look, I have come to do Your will." He cancels the first covenant in order to put the second into effect. (Hebrews 10:9 NLT)

## CHRISTIANS AND THE SABBATH

No one is required to keep the Sabbath today. However, some traditional Jewish people are still attempting to live by the Mosaic laws of the Old Testament, including the Sabbath.

In the following verses, followers of Christ are told not to let others condemn them of sin for not keeping the Sabbath:

> So don't let anyone condemn you for what you eat or drink, or for not celebrating certain holy days or new moon ceremonies or Sabbaths. For these rules are only shadows of the reality yet to come. And Christ Himself is that reality. (Colossians 2:16–17 NLT)

> One person regards one day holier than other days, and another regards them all alike. Each must be fully convinced in his own mind. The one who observes the day does it for the Lord. (Romans 14:5–6 NET)

The first converts to Christianity were Jewish people. Some continued to obey the Old Testament commandments, including meeting on the Sabbath (Saturday). They would also meet on Sundays to celebrate the Lord's Day (his resurrection day). However, some also felt that Gentile Christians should be required to obey the Old Testament commandments.

> While Paul and Barnabas were at Antioch of Syria, some men from Judea arrived and began to teach the believers: "Unless you are circumcised as required by the law of Moses, you cannot be saved." Paul and Barnabas disagreed with them, arguing vehemently. Finally, the church decided to send Paul and Barnabas to Jerusalem, accompanied by some local believers, to talk to the apostles and elders about this question. (Acts 15:1–2 NLT)

Paul and Barnabas took the issue to the leaders of the new Christian religion in Jerusalem, known as the Jerusalem Council.

Following is the Council's decision concerning this critical issue:

> "For it seemed good to the Holy Spirit and to us to lay no greater burden on you than these few requirements: *You must abstain from eating food offered to idols, from consuming blood or the meat of strangled animals, and from sexual immorality.* If you do this, you will do well. Farewell." The messengers went at once to Antioch, where they called a general meeting of the believers and delivered the letter. And there was great joy throughout the church that day as they read this encouraging message. (Acts 15:28–31, author's emphasis)

Important: The only requirements for Gentile (and all) Christians were to abstain from pagan practices that would cause non-believers to think they were still pagans. This also applied to the nation of Israel and the Jewish converts to Christianity. Therefore, Gentile Christians were not required to keep the Jewish Sabbath.

## ETERNAL REST FOR GOD'S PEOPLE

The Father God promises his children they can look forward to spending eternity with him. It will be an eternity where they will never grow weary and need to rest their bodies or minds.

> Consequently a Sabbath rest remains for the people of God. For the one who enters God's rest has also rested from his works, just as God did from his own works. Thus we must make every effort to enter that rest, so that no one may fall by following the same pattern of disobedience. (Hebrews 4:9–11 NET)

There will be no commandments to keep in order to please the Father God in Heaven or on the new Earth. We will see the Father and Jesus face-to-face and enjoy their companionship. We will continually praise and worship them without growing weary and needing rest.

## CHAPTER QUESTIONS

DQ1: What reason does God state in Genesis 2:2–3 for why he rested from his work of creation?

DQ2: From the scriptures, who is commanded to keep the Sabbath (Saturday), and what is God's reason for this commandment?

DQ3: Why was there a need to replace the Old Covenant with the New Covenant?

DQ4: How did this impact the requirement to keep the Sabbath?

## DIG DEEP

DD1: The death and resurrection of Jesus Christ were essential to the establishment of the New Covenant. Why do the people of Israel 2,000 years later still refuse to accept Jesus as the Messiah who established the promised New Covenant?

## PERSONAL APPLICATION

PA1: Why is it important for you to rest one day a week?

# PART E

Other Creations

# 24.

## What Else Was Designed and Created?

Even though Jesus Christ finished his creation of the heavens, Earth, and all living things described in Genesis chapters 1 and 2, he is not limited in his ability to create whatever and whenever the Father desires. The following would not exist without our Creator God.

### INVISIBLE REALM AND ANGELS

In the following verses, we read the natural world *(visible)* and the supernatural world *(invisible)* were created by Jesus.

> He is the image of the invisible God, the firstborn of all creation. For by Him all things were created, both in the heavens and on earth, *visible and invisible*, whether *thrones or dominions or rulers or authorities—all things* have been created through Him and for Him. He is before all things, and in Him all things hold together. (Colossians 1:15-17 NASB, author's emphasis)

We know from our earlier study on angels that they are part of the supernatural realm designed by the Father and created by Jesus. We see this again in the above verses since *thrones or dominions or rulers or authorities* refer to various levels of angels. (We know that the Father God, Jesus Christ, and the Holy Spirit were not created but are part of the supernatural realm.)

### MELCHIZEDEK

Melchizedek is a mysterious figure described in the Old Testament (Genesis chapter 14) and the New Testament book of Hebrews.

Following are some verses about him in the Book of Hebrews:

> Now this Melchizedek, king of Salem, priest of the most high God, met Abraham as he was returning from defeating the kings and blessed him. To him also Abraham apportioned a tithe of everything. His name first means king of righteousness, then

king of Salem, that is, king of peace. Without father, without mother, without genealogy, he *has neither beginning of days nor end of life* but *is like the son of God*, and he remains a priest for all time. (Hebrews 7:1–3 NET, author's emphasis)

Mystery: Do these verses imply that Melchizedek was an eternal being who was not created? (Jesus was not created.) Or, as some say, do they mean that he was a created human being who lived in Old Testament times, but no one knew his family lineage?

Instructional Comment: Like all mysteries in the Bible, we do not have enough biblical information to know whether Melchizedek was a supernatural being or simply a great human being without a known human lineage.

## BELIEVERS ARE A NEW CREATION IN CHRIST

A significant change occurs when a person is born-again. They become a *new creation*. The Holy Spirit gives them a brand-new nature, one that never previously existed in them. This supernatural new nature exists alongside their old nature.

Therefore, if anyone is in Christ, the new creation has come. (2 Corinthians 5:17 NIV)

The Greek word for *new* is *kainos* and means it's a nature that didn't previously exist.

### Unique New Nature

The Holy Spirit doesn't take the existing self-centered old nature, modify it, improve it, or renovate it to make it better. If you are a follower of Jesus Christ, you can't live the new Christian life in a way that pleases God with just an improved, self-centered old nature. You must be given a new supernatural spiritual nature, enabling you to experience a new spiritual relationship and life with the God of the Bible.

Think about it this way. You have a forty-year-old home you decide to improve. You renovate it so it looks and functions much better than it did before. However, its foundation and framework are still the same old construction. Later, you decide to build a brand-new home. It's a new construction that previously never existed. Your new nature is like this newly constructed home.

## CHRISTIANITY AND THE CHRISTIAN CHURCH

Christianity and the Christian church did not exist until its birth on the Pentecost, as described in Acts 2:1–22. The Christian church is not a physical building or an assembly of people meeting in one. Instead, it is supernatural in the form of every born-again follower of Jesus Christ in Heaven and on Earth. Together, we are called the body of Christ.

The human body has many parts, but the many parts make up one whole body. So it is with the body of Christ. (1 Corinthians 12:12 NLT)

Christ is also the head of the church, which is His body. (Colossians 1:18 NLT)

When the Holy Spirit birthed the Christian church, the Father put his Son, Jesus Christ, in charge of it. This is because the followers of Christ are the members of his supernatural body.

## HEAVEN

The Father God, the Intelligent Designer of everything, designed and built Heaven as a city with permanent foundations for his throne and residence.

> Abraham was confidently looking forward to a city with eternal foundations, a city designed and built by God. (Hebrews 11:10 NLT)

Note: When the physical body of followers of Jesus Christ dies, their eternal being of a spirit/soul goes immediately to Heaven.

## HELL/HADES/SHEOL

When the physical body of those who reject Jesus Christ dies, their eternal being of a spirit/soul goes immediately to Hell. There are other names for this place with slightly different perspectives, but they all represent the same place. Jesus used Hell and Hades interchangeably, so we can assume they are the same place. The Apostle Peter quoted an Old Testament verse with the Hebrew word Sheol and changed it to Hades in the New Testament.

Note: Hell/Hades/Sheol must have been designed and built for their purposes, just as Heaven was designed and built for its purpose.

## GARDEN OF EDEN

In Genesis 2:4–9 we read about the creation of the Garden of Eden on Earth.

> This is the account of the heavens and the earth when they were created—when the LORD God made the earth and heavens. Now no shrub of the field had yet grown on the earth, and no plant of the field had yet sprouted, for the LORD God had not caused it to rain on the earth, and there was no man to cultivate the ground. Springs would well up from the earth and water the whole surface of the ground. *The LORD God formed the man from the soil of the ground* and breathed into his nostrils the breath of life, and the man became a living being. *The LORD God planted an orchard in the east, in Eden; and there he placed the man he had formed. The LORD God made all kinds of trees grow from the soil*, every tree that was pleasing to look at and good for food. (Now the tree of life and the tree of the knowledge of good and evil were in the middle of the orchard.) (Genesis 2:4–9 NET, author's emphasis)

Notice that Adam was created before the Garden. Like Adam, the Garden was made from what had already been created (natural elements of the Periodic Table).

Mystery: How long was it from when Adam was created to when the Garden of Eden was made, and how long was it before the trees produced fruit for Adam and Eve to eat?

# MARRIAGE

The Father God designed, and Jesus created the marriage relationship between a husband and wife with Adam and Eve in the Garden of Eden. Adam was made from the elements of the Earth, and Eve was made from Adam's rib bone to demonstrate their unique interdependence upon each other.

> So the man named all the animals, the birds of the air, and the living creatures of the field, but for Adam no companion who corresponded to him was found. So the LORD God caused the man to fall into a deep sleep, and while he was asleep, he took part of the man's side and closed up the place with flesh. Then the LORD God made a woman from the part he had taken out of the man, and he brought her to the man. Then the man said, "This one at last is bone of my bones and flesh of my flesh; this one will be called 'woman,' for she was taken out of man." That is why a man leaves his father and mother and unites with his wife, and they become a new family. (Genesis 2:20–24 NET)

# DIVERSITY OF PEOPLE

No Diversity Before the Flood

Before the flood, everyone on Earth lived in the same geographic area, spoke the same language, and had the same culture.

> Now the whole earth had one language and the same words. (Genesis 11:1 ESV)

## No Diversity After Flood

Noah and his family would have spoken the same language and shared the same culture as their ancestors. The entire Earth's population later originated from these eight people on the ark. (The ark is assumed to have settled on Mount Ararat in modern-day Turkey.)

> The sons of Noah who came out of the boat with their father were Shem, Ham, and Japheth. (Ham is the father of Canaan.) From these three sons of Noah came all the people who now populate the earth. (Genesis 9:18–19 NLT)

## Diversity Only After Tower of Babel

Noah's descendants founded several cities in Mesopotamia, such as Babylon, Calah, Erech, and Nineveh (Genesis 10:10–31) before they built the Tower of Babel.

We see in the following verses that God scattered the people worldwide at the Tower of Babel:

> And the LORD said, "Behold, they are one people, and they have all one language, and this is only the beginning of what they will do. And nothing that they propose to do will now be impossible for them. Come, let us go down and there confuse their language, so that they may not understand one another's speech." So the LORD dispersed them from there over the face of all the earth, and they left off building the city. Therefore its name was called Babel, because there the LORD confused the language of all the earth. And from there the LORD dispersed them over the face of all the earth. (Genesis 11:6–9 ESV)

When he did this, the scattered people developed their own languages and cultures. A globally diverse people developed over time. Different physical traits developed along with genetic adaptations. This was the result of humankind's microevolution, the human species' adaption to their new environments.

Mystery: What language did Adam and Eve, Noah, and his family speak?

Instructional Comment: We can't know what language Adam and Eve spoke. Moses wrote the book of Genesis primarily in ancient Hebrew. Because he wrote in the Hebrew language, we cannot assume Adam and Eve spoke Hebrew. According to Jewish tradition, God gave them the "Adamic language," which he used to communicate with them. Some Bible scholars believe they spoke a unique language that has been lost over time. A friend of mine believes they spoke the language of the angels (1 Corinthians 13:1) since it is a pure language not of this Earth. It's interesting to speculate sometimes to see what alternatives we can develop about mysteries. But we must be careful and not promote our speculations as facts.

## NATION OF ISRAEL

The Father God called Abram out of the city of Ur in the land of the Chaldeans (modern-day Iraq) to become the father of the people and nation of Israel.

---

> God desired to have a family, so he created the nation of Israel to be his children. After the death and resurrection of Jesus Christ, born-again followers of Jesus become children of God.

---

> Now the LORD said to Abram, "Go out from your country, your relatives, and your father's household to the land that I will show you. Then I will make you into a great nation, and I will bless you, and I will make your name great, so that you will exemplify divine blessing." (Genesis 12:1–2 NET)

> The LORD said to him, "I am the LORD who brought you out from Ur of the Chaldeans to give you this land to possess." (Genesis 15:7 NET)

### The Law and the Prophets of Israel

The Father God designed the Mosaic Law to provide structure, civil order and governance, moral codes, and ceremonial forms of worship for the nation of Israel.

> Observe the requirements of the LORD your God, and follow all His ways. Keep the decrees, commands, regulations, and laws written in the Law of Moses so that you will be successful in all you do and wherever you go. (1 Kings 2:3 NLT)

He also designed the role of prophets to speak for him to the people of Israel:

> Now the LORD spoke through His servants the prophets. (2 Kings 21:10 NASB)

These gave the people of Israel warnings and instructions, including how to treat each other, as Matthew stated in the New Testament:

> In everything, therefore, treat people the same way you want them to treat you, for this is the Law and the Prophets. (Matthew 7:12 NASB)

Instructional Comment: Even though Jesus and the Holy Spirit are mentioned many times in the Old Testament scriptures, the Jewish people did not understand the idea of a Triune God of the Father, his Son, Jesus, and the Holy Spirit. When you see "God" and "LORD" in Old Testament scriptures, the Jewish people think of the Father God only.

## UNITED STATES OF AMERICA?

The answer to the question, "Was America founded as a Christian nation by the Creator God," hinges on understanding the word "founded." This word has caused controversy among people and scholars who have attempted to determine whether the origin of the United States of America was Christian. Was America founded as a Christian nation over 400 years ago when the first Christian settlers came? Or more than 150 years later, when the "founding fathers" formed the thirteen colonies into an independent nation with its civil government?

Sometimes, the focus on the people (and dates) who founded America can get in the way of the more important questions, "Was America created by the Father God, Jesus Christ, and the Holy Spirit?" "And did they create it to be a Christian nation?" Let's take a quick look at early American history to see if we can answer these questions.

### First American Christian Settlement

We need to consider the religious attitudes and values of the initial European Protestant settlers. Many of these settlers were serious Christians who established laws, constitutions, and practices reflecting strong Christian values for the eventual thirteen colonies that became the United States of America.

The first successful European colony in America was established in 1607 in Jamestown, Virginia. In 1620, New England was settled by a diverse group of people, many of which were Puritans fleeing religious persecution. These Pilgrims sailed on the Mayflower headed toward Virginia but were knocked off course by a storm and landed in Plymouth, Massachusetts. Before landing, they wrote the Mayflower Compact. Its purpose was to define laws for civil government and social order. It was an agreement that the colonists would live according to the Christian faith.

## Declaration of Independence and God

America was founded on biblical principles. Such things as freedom, justice, individual rights, and more come from the Bible and the colonists' oppressive experiences with the British King (George III).

The following is from the second paragraph of the Declaration of Independence, signed in 1776:

> We hold these truths to be self-evident, that all men are created equal, that they are endowed by their Creator with certain unalienable Rights, that among these are Life, Liberty and the pursuit of Happiness—That to secure these rights, Governments are instituted among Men, deriving their just powers from the consent of the governed—That whenever any Form of Government becomes destructive of these ends, it is the Right of the People to alter or to abolish it, and to institute new Government, laying its foundation on such principles and organizing its powers in such form, as to them shall seem most likely to effect their Safety and Happiness.

## America's Constitution

Most people and scholars agree that America was founded in 1789 as an independently governed nation with the ratification of our Consitution (after the Declaration of Independence in 1776). Fifty-two of the fifty-five signers of the Declaration of Independence stated they were Christian. Some people today wonder what they meant when they said they were Christian. Were they devoted followers of Jesus Christ who worshipped him, prayed, attended church, and did all they could to promote the cause for Christ in our newly founded nation? Or were they Christians in name only? Perhaps some were Deists who believed in a supreme creator god based on reason, not the Almighty God of the Bible. We don't know what was in their hearts, but from various things they wrote and their quotes, it seems that most were followers of Jesus Christ.

Note: The Declaration of Independence and the U.S. Constitution are in the public domain so that they can be downloaded, printed, reproduced, and used without permission.

## CHAPTER QUESTIONS

DQ1: What would life be like if there was no supernatural realm of angels and demons?

DQ2: Why is the biblical Christian church not a physical building?

DQ3: What was God's ultimate purpose in dispersing people worldwide?

DQ4: From the Scriptures, what were God's reasons for creating a new nation (Israel) that had never existed?

## DIG DEEP

DD1: What is the likely reason many Christians believe Melchizedek was a created human being who lived in the Old Testament times rather than an eternal being like Jesus?

## PERSONAL APPLICATION

PA1: How does it change your feelings about America when you consider it might have been created by God for His purposes?

# PART F

## The Theory of Evolution

# 25.

## Evolution of Species Basics

This chapter discusses some basic theories concerning how biological organisms are said to have evolved over millions and billions of years on Earth. *I refer to this as the biological evolution of species theory, which speculates that one species evolves into another different species.*

### WHY STUDY EVOLUTION?

Let's start with a question: "Why do born-again followers of Jesus Christ need to study and learn about anything not in the Bible?" After all, we have been taught everything needed to know about God, salvation in Christ, and how to live in the Bible (especially the New Testament).

I have been an ardent follower of Jesus Christ for nearly fifty years. When I started my lifelong journey with the God of the Bible, I did so, choosing to know nothing about him except what was in the Bible. I wanted to be anchored in the truths of the Bible alone. I did not read a single book about Christianity until some years later. I wanted to know and use the word of God accurately. Since then, however, I have read many books on biblical and non-biblical topics.

> Do your best to present yourself to God as one approved, a worker who has no need to be ashamed, rightly handling the word of truth. (2 Timothy 2:15 NASB)

### Knowing Other Subjects about Life

Reading only the Bible after being anchored in its truths limits a Christian's ability to understand the myriad of issues of life. When I hear well-meaning Christians say they don't read anything except the Bible, I know they are either young Christians (as I was) who still need to learn the basics of the Christian faith (1 Corinthians 3:1–3) or are misinformed. Life includes so many issues that are not covered in the Bible. I believe followers of Jesus Christ should be well-informed about things they will encounter in life. We certainly cannot anticipate everything we need know or want to know beforehand. And we are bombarded by information overload every day. So, how do we choose a subject to learn about?

Following is my wife's and my story about this:

> My wife and I love to research new things. We are not indiscriminate about this. We pray about what God would have us do each day, and the Holy Spirit leads us. We know we do much more research (including the Bible) than many Christians and people. That's just part of the Creator's purpose for us. It's part of who we are. But what about you? If you do not have this thirst to know more, what and how often should you seek new knowledge outside the Bible? Only you and the Spirit of God can determine this. I suggest you pray each morning before you start your day and ask the Holy Spirit to guide and teach you what he wants for you and for others you will encounter that day (John 14:26).

## What about Evolution?

Why should Christians study evolution? Why might it be important for your knowledge and help you converse with others? Evolution is a topic many people know little about, so arm yourselves with the truths of the Bible and the sciences to understand the unproven theories of evolution. Who knows when you might have an encounter that will enable you to have an impact and help change a person's heart and life?

---

Evolution is a topic many people know little about. It's important to learn the truths of the Bible and the sciences to understand why creation is valid and evolution is not.

---

Following is a true story about how there might have been a different outcome if I had understood creation and evolution well. It might have been different if I had known how to use Christian biblical apologetics about creation.

> I was a new believer in Christ and was at a party at a friend's apartment. I met another college-age young man, and we started the typical conversation about who you are and so on. At one point, I told him I was a Christian and asked if he was. He said he wasn't and would never become one. I asked why, and he said he was a biology major in college and believed in evolution and not creation. He said evolution was proven truth and creation was unproven fiction. I did not know anything about evolution, and I only knew a little about creation. So we just argued about the things neither of us knew much about. The result was he went away even more convinced that Christians were ignorant about things of the world they should know about.

If I had known then what I know now about creation and evolution, perhaps this young man would be in the Kingdom of God. I can't know this for certain, but he would have at least heard facts about creation and evolution that could disprove his theories. God's living word (Hebrews 4:12–13) equips us to respond with truth and grace when needed (2 Timothy 3:16–17).

## BASIC THEORY OF BIOLOGICAL EVOLUTION OF SPECIES

The foundational supposition of the theory of biological evolution is that accidents of nature result in biological and genetic changes within a species over long periods, enabling it to eventually evolve into a new species. (This is also known as macroevolution.)

### No Evidence Found to Support the Theory

The basic theory of the biological evolution of species says that changes eventually lead to the formation of a new species (or genus) from the prior species. However, it should be noted that fossil records, field observations of current species, and laboratory experiments have never proved this dramatic change from one species to a new species. Instead, based on their observed evidence, evolutionists infer how a species might have evolved physically, behaviorally, and genetically over long periods. It's also important to note that evolutionary scientists can only study physical evidence for their theories. Physical evidence cannot tell us anything about the subjective nature of the soul. Therefore, no physical evidence in the natural world can scientifically support the existence of the sentient nature of complex living organisms such as human beings.

## BASIC THEORY OF HUMAN EVOLUTION

For human beings, the theory of biological evolution says that people have gradually evolved to be "human" over several billion years. They say modern people developed physical traits and behaviors over billions of years from ancient ancestors, including primates and other living organisms from which they descended. But they also say modern people did not descend directly from any *living* primate or monkey. Instead, they promote the idea that people share a common ape ancestor with modern chimpanzees who lived between eight and six million years ago. (Periods to evolve differ between evolution websites.) These scientists say that humans and chimpanzees evolved along different evolutionary paths from that common ape ancestor. This theory of direct descent was the foundation on which Darwin and other early evolutionists built their theories of the evolution of species.

## DARWIN'S THEORY OF EVOLUTION BY NATURAL SELECTION

In 1859, Charles Darwin published his book, *On the Origin of Species*. His theory is known as the Theory of Evolution by Natural Selection. Through natural selection, his theory says that diverse species could evolve from a common ancestor. (This is direct descent.) He proposed that living organisms change over long periods as they adapt to their environment. Individuals that adapt well survive and produce offspring that will survive. (Later, some people referred to this as the "survival of the fittest.") As the offspring reproduce, more of the species' population will have these new traits. Once the new traits are widespread throughout the population, the species' population is said to have evolved into a new species.

Interestingly, Darwin originally believed in the Genesis account of creation orchestrated by a "designer." It wasn't until he visited the Galápagos Islands that he began to rethink his view of the natural, biological world. As he visited various islands in the chain, he began to notice what appeared to be variations of a species. It seemed to him that evolution was occurring independently on each island as the species adapted to the unique environment on the island. He noticed this in South American countries as well. This caused Darwin to conclude that new species evolve from previous species over significant periods through natural selection. *The basis of his theory is that some better-adapted species variations could eventually survive and reproduce in their natural habitat to become a new species.*

Website: You can read more about Charles Darwin and his theories about evolution and natural selection on the following Smithsonian website: https://www.smithsonianmag.com/science-nature/the-evolution-of-charles-darwin-110234034

Reminder: Darwin's ideas about evolution are still scientifically unproven theories.

## CHARLES DARWIN DID NOT KNOW ABOUT GENETICS

Charles Darwin did not know how traits were passed on to offspring. He just observed that this occurred. He did not know about genetics and how genes encode certain traits that are passed on to future generations. He also did not know about genetic mutations and how they can affect changes in physical and behavioral characteristics in species.

Note: Charles Darwin was a Naturalist who studied and theorized about the biological, physical world of nature. His theories did not extend beyond planet Earth into the universe, other than saying the Earth was likely older than the commonly thought 6,000 years.

Definition: *Macroevolution* is the theoretical process by which ancient species change and evolve into modern species. Early theories assumed it was a direct descent process (which many scientists now reject).

*Natural selection* is a mechanism that can result in the microevolution of species (not macroevolution). It's the process by which species better suited for the environment successfully survive and reproduce.

*Adaptations* are physical and behavioral traits that make a species better suited to its environment to survive and potentially change. Adaptations are, therefore, the engine that drives natural selection for microevolution.

## MACROEVOLUTION

Macroevolution is the theory of biological evolution of species. It is the theory that a long evolutionary process occurs by which species undergo significant genetic and trait changes at or above the species level in the biological Tree of Life. (This is not the biblical Tree of Life in the Garden and the new Earth.) Science tells us that for macroevolution to occur, massive amounts of new genetic material and related information must be created in a species for a new species to evolve. This new genetic material has never been found to exist in a species.

Important: There are no examples of macroevolution to prove that massive amounts of new genetic material have ever been created within a species.

Website: More information about this view of macroevolution may be found on the following website: https://evolution.berkeley.edu/evolution-101/macroevolution/what-is-macroevolution/

## MISSING LINKS

The macroevolution theory requires transitional species, also known as "missing links." A missing link is a vacant placeholder to fill a gap in the evolutionary tree for a non-existent transitional species. It's a gap in the worldwide fossil records for a species to prove how it evolved into a different species over time. For example, for vertebrates to move out of the water onto land to walk, there must have been an initial ancient sea creature that developed lungs to breathe air and feet to walk on land. This transitional species is a missing link because no fossil remains have been found to support the theory. In other words, *missing links do not exist.*

### Proposed Human Missing Links

Two popular discoveries were declared a missing link between apes and humans. Later, both were proved to be wrong. The Piltdown Man, *Eoanthropus dawsoni*, in December 1912, was said to be a missing link between apes and humans that would have lived half a million years ago. Children were taught that the Piltdown Man was a missing link for over forty years. It wasn't until 1953 that it was revealed to be a fraud. It was simply a human skull with orangutan teeth that had been filed to look human.

### "Lucy" Was Not a Missing Link

"Lucy" is the nickname of skeletal remains (40 percent complete) discovered in 1974 in Ethiopia, Africa. Donald Johansen, who made the discovery, declared this fossil the missing link between apes and modern humans. The scientific name of Australopithecus afarensis was given to the fossil nicknamed Lucy. Many school textbooks erroneously state that Lucy was our ancient ancestor with human-like features.

Evolution scientists state that Lucy was a missing link because they proposed she had both human-like and distinct ape-like features. However, subsequent closer, non-biased examinations of Lucy's skeletal features identified her as an ape, not a human missing link. For example, Trey Bowling with Dr. Brian Thomas, "The Soulless Hominid Theory: A Fatal Flaw in Old Earth Creationism." (Website: https://www.icr.org/article/15206.) Also, scientists Jack Stern and Randall Sussman analyzed Lucy's skeleton and reported in the Journal of Physical Anthropology that Lucy had many ape-like features. ("The locomotor anatomy of Australopithecus afarensis.") A critical finding was that Lucy's pelvic bone was not made for walking upright (bipedal) as an ancient human ancestor would need. It was, in fact, ape-like, resembling that of a chimpanzee. Dr. Charles Oxnard, a professor of anatomy, is an expert on the Australopithecus afarensis fossils. In his book, *The Order of Man: A Biomathematical Anatomy of Primates*, he states that no Australopithecine fossils have been proven to be bipedal. Therefore, Lucy could not have been a bipedal human ancestor of modern humans.

Website: You can read more about Lucy and why she is not the missing link between apes and people on this creation website: https://genesisapologetics.com/creation-vs-evolution-fast-facts

Note: Despite this, many evolutionary scientists still accept Lucy as a valid missing link in their human evolution lineage models.

### Missing Link Term Deemphasized

Because of discoveries and resultant analyses like that of Lucy, evolutionary scientists today prefer not to use the term "missing link." This is because it has not and cannot be proven. Many scientists use other methods to theorize the evolution of modern species from ancient ones. Modern evolutionists have shifted from this cornerstone belief to one that does not require proof of direct descent human lineage.

## NO MISSING LINKS IN THE BIBLE

Nothing in the Bible indicates or implies the existence of missing links (transitional species). The Triune God did not make one species and cause or allow it to evolve into different and separate species over millions or billions of years. For example, when he created domestic cats, he made a creation kind, a master species of the feline family with many variations. There are Burmese, Maine Coon, and Persian cat variants of the same domestic cat. Minor changes in the domestic cat species have occurred over time due to adaptation to their geography and environment. This is known as "microevolution," which all Christians should be aware of and accept as scientifically true. Both evolutionary and creation scientists accept the truth of microevolution.

Important: Nothing is ever created by unguided, random, natural processes. Nothing cannot create something.

## CHAPTER QUESTIONS

DQ1: Why do followers of Jesus Christ need to study and learn about the theories of biological evolution?

DQ2: Briefly define Darwins' theory of biological evolution.

DQ3: How can the soul, composed of thought, will, emotions, and decision-making, evolve from the material, natural world where the soul does not exist?

DQ4: With only physical evidence to rely on, how does an evolutionist answer the following question many people have, "What is the purpose of my existence?"

## DIG DEEP

DD1: Listen to the podcast by Trey Bowling with Dr. Brian Thomas, "The Soulless Hominid Theory: A Fatal Flaw in Old Earth Creationism." Explain why Dr. Thomas says the Australopithecus africanus skeleton named "Lucy" was not a transitional species to modern humans. Website: https://www.icr.org/article/15206.

## PERSONAL APPLICATION

PA1: What have you heard about missing links and why they cannot answer questions about life, such as, "Why am I here?"

# PART G

Delving Deeper
into Theories of Evolution

# 26.

## Delving Deeper: More Theories of Evolution (I)

Important: Delving Deeper chapters contain material that may be difficult to understand. I suggest you spend extra time studying it and ask for the Holy Spirit's help. The Father and Jesus sent him to live within Christians as their Teacher, Guide, and Helper.

Note: Studying the evolution of species will reveal many variations of the theory and how it works in nature. These variations indicate why it's still a theory and not a proven scientific fact.

### LAMARCK: FIRST FULL THEORY OF EVOLUTION BY NATURAL SELECTION

In the 1700s, scientists began speculating that life was not created. French naturalist Jean Baptiste Pierre Antoine de Monet, Chevalier de Lamarck, in 1801, was the first to propose that species changed over long periods into new species. Prominent in his theory are the words "use" and "disuse" of organs (heart, lungs, feet, and so on). He theorized that as environments changed, organisms that adapted and changed their traits and behavior survived. They would survive when they used (use theory) an organ to adapt more than they needed its use in the past. It would become a more significant part of the organism. Organs not conducive to survival would be used less, eventually shrinking and becoming less significant (disuse theory). He believed that as they adapted to their environments, this use and disuse in the process of natural selection caused them to evolve from simple to complex living forms. Therefore, he felt God wasn't needed to create an organism or assist in its evolution. They just evolved on their own without outside assistance.

## MODERN SYNTHESIS (NEO-DARWINIAN EVOLUTION)

This is an updated version of Charles Darwin's Theory of Evolution by Natural Selection that embeds the study of genetics and population changes. It is another form of macroevolution.

Essentially, Modern Synthesis proposes that:

- Variations in genes create changes in physical traits (phenotype)
- Trait variations are due to randomly occurring gene mutations affecting the organism's survival
- Natural selection is the result of adaptation to the environment
- The accumulation of selected genetic mutations over time results in the evolution of the population
- Evolution occurs at the population level, resulting in a new species

Modern Synthesis is a commonly accepted theory of biological evolution among scientists. However, some scientists disagree in part or totally with the core principles mentioned above.

Website: The following website has detailed information about Modern Synthesis: https://darwin200.christs.cam.ac.uk/modern-synthesis

## SPECIATION

One way to define speciation is to say it is an evolutionary process that leads to the formation of new, genetically distinct species reproductively isolated from one another. Evolutionary scientists think significant geographical changes occur that separate the species into two or more geographic populations. As a result, organisms from these separated populations can no longer access one another for breeding. Over long periods, the populations better adapt to their environment through natural selection and beneficial gene mutations that lead to genetically and physically distinct species. Examples of events that can result in geographic barriers to breeding include volcanoes forming mountains, deep valleys, or islands; formation and movement of glaciers; new river formations; or habitat separation created by human activity. Also, the unintentional movement of a species to isolated geographic areas can occur through transportation, often on ships, resulting in reproductive isolation. It is another variation of the theory of macroevolution.

Website: More information on evolutionary speciation can be read on the following website: https://biologydictionary.net/speciation

## MICROEVOLUTION

Microevolution is the process of adaptation by which changes to a species' genetic makeup and physical traits occur in small ways over short periods. Many creationists accept this theory of environmental adaptation within a species. Some evolutionists theorize that microevolution is the driving mechanism for the macroevolution process over billions of years. This, too, has never been proven scientifically to be true.

---

> Worldview: Microevolution and adaptation do not prove macroevolution, which is the theory of biological evolution.

---

## ADAPTATION AND MICROEVOLUTION IN HUMANS

Microevolutionary changes within the human species can enable people to survive and thrive within a particular geographic environment. Let's read about the Tower of Babel in the Bible to understand how this adaptation to different geographic environments might have resulted in the many variations of the human species we see around the world today:

> At one time all the people of the world spoke the same language and used the same words. As the people migrated to the east, they found a plain in the land of Babylonia and settled there. They began saying to each other, "Let's make bricks and harden them with fire." (In this region bricks were used instead of stone, and tar was used for mortar.) Then they said, "Come, let's build a great city for ourselves with a tower that reaches into the sky. This will make us famous and keep us from being scattered all over the world." But the LORD came down to look at the city and the tower the people were building. "Look!" He said. "The people are united, and they all speak the same language. After this, nothing they set out to do will be impossible for them! Come, let's go down and confuse the people with different languages. Then they won't be able to understand each other." In that way, the LORD scattered them all over the world, and they stopped building the city. That is why the city was called Babel, because that is where the LORD confused the people with different languages. In this way He scattered them all over the world. (Genesis 11:1–9 NLT)

Note: People have continued to adapt to their environments for thousands of years since the Tower of Babel. World population estimates indicate that there are nearly eight billion people spread out all over the world in many different geographies and climates. Just think of the adaptation changes seen in people groups in Scandinavian, African, and Asian countries.

## CHAPTER QUESTIONS

DQ1: Describe how the Modern Synthesis theory differs from Darwin's theory of biological evolution of species.

DQ2: Describe how the Speciation theory differs from the Modern Synthesis theory.

DQ3: Why should creationists accept microevolution as valid?

DQ4: Explain why the biblical story of the Tower of Babel exemplifies microevolution for human beings.

## DIG DEEP

DD1: Explain why Lamarck's "use" and "misuse" theory is an example of the theory of biological evolution.

## PERSONAL APPLICATION

PA1: In what ways has this material on the biological evolution of species changed your perspectives on the God of the Bible?

# 27.

# Delving Deeper: More Theories of Evolution (II)

Important: Delving Deeper chapters contain material that may be difficult to understand. I suggest you spend extra time studying it and ask for the Holy Spirit's help. The Father and Jesus sent him to live within Christians as their Teacher, Guide, and Helper.

## SCIENTIFIC THEORY OF INTELLIGENT DESIGN

I was surprised when I read about this theory that proposes that the complexity and intricate design of the universe and living organisms indicate an intelligent cause (design) rather than random variations. This theory was developed from observations and inferences from the natural world. The theory does not challenge the traditional Darwinian idea of a long evolution process. However, it challenges the Darwinian theory that undirected processes such as natural selection acting on random variations can cause the intricately complex designs observable in nature. These scientists believe that intelligent design can occur naturally without an Intelligent Designer (as Christians believe). They propose that God is not required to create the naturally occurring intelligent design we see in such things as the digital code in DNA, "miniature machines" in cells, and the fine-tuning seen in the laws and constants of physics. Some of these scientists suggest that a guiding intelligence was involved in the intelligent design of the universe and all living organisms. However, they are not willing to admit that their idea of an intelligent designer is the Creator God of the Bible.

Website: More information about the scientific theory of intelligent design can be found on the following website: https://intelligentdesign.org/whatisid

Reminder: The *Intelligent Designer* is the *Father God* who decided how he wanted creation to exist (1 Corinthians 8:6; Hebrews 11:10). *Jesus Christ* is the person of God who carried out the Father's design when he spoke creation into existence (Colossians 1:15–17). He also sustains creation according to the Father's design and intended purposes. The *Holy Spirit* is the person of God who has breathed the breath of life into people (Job 33:4).

## EVOLUTIONARY CREATIONISM

This theory embraces the biblical idea of creation. However, it proposes that God created everything through a designed, orderly, gradual process of evolution over billions of years. It says that modern people evolved from pre-human ancestors over a lengthy period. During this time, the image of God and human sin gradually developed in people. Individuals who believe this refer to themselves as Christians because they love God and strive to live for him. They say they have a personal relationship with Jesus Christ and live by the power of the Holy Spirit. They say they experience the supernatural work of the Holy Spirit in their lives.

Consider these questions about the biblical accuracy of this theory:

- How can the image of God and human sin gradually develop in people? Does this view eliminate the rebellion of Adam and Eve in the Garden of Eden? (That's when biblical sin entered the world.)
- At what point in the evolution of people did God say, "OK, people have learned enough about me, so now they have my image?"
- How could people made in God's image have a sinful nature unless they rebelled against him?
- If they rebelled, why would he give them his image and holy nature?
- Weren't Adam and Eve sinless until they rebelled against God in the Garden of Eden?
- Isn't this theory contrary to the biblical account that Adam and Eve were the only people initially created in God's image (Genesis 1:26–28)?

Website: More information about Evolutionary Creationism can be found on the following website: https://www.scienceandfaith.org/evolutionary-creationism

### Christians without Old Testament Knowledge?

It seems that people who believe in Evolutionary Creationism don't have a solid biblical understanding of creation. However, does an inaccurate knowledge of the Old Testament of the Bible prohibit someone from believing in the death and resurrection of Jesus Christ and being saved? I know many born-again followers of Jesus Christ who do not understand the Old Testament of the Bible. But I am certain they have committed their lives to Jesus and are striving to live a Christian life that honors him. Perhaps some people who profess Evolutionary Creationism are Christians. God knows what's in their hearts.

## THEISTIC EVOLUTION

The beliefs of Theistic Evolutionists are similar to Evolutionary Creationists in the following ways:

- Jesus created everything, including the universe
- Some of the theories of evolution are valid
- There was intelligent design in creation

I also see differences between them. Maybe the most important differences are the following:

- Theistic Evolutionists focus primarily on the universe, its origin, and its evolution
- Theistic Evolutionists believe Jesus created the singularity substance and then kicked off the evolution of the universe, allowing it to evolve on its own without his interaction
- Evolutionary Creationists seem to focus mainly on the biological world, its origin, and the evolution of species

Even with the differences, some people consider these theories to be the same belief system.

## BIOLOGICAL TREE OF LIFE THEORY

Biology's Tree of Life is also called a "phylogenetic tree." It depicts the theoretical evolutionary relationships for organisms as they evolved from their beginning into current species. In this theory, each upward branch on the tree is a new species that evolved over millions or billions of years from a previous species. Charles Darwin published *On the Origin of Species* in 1859 and postulated the first Tree of Life as a simple structure. The Tree has been modified multiple times over the decades. It is based on the assumption that all life evolved from chemicals in the Earth's soil (primordial soup) that combined to form one-cell organisms. These one-cell organisms merged and replicated to form multicell living organisms. This process continued over billions of years to produce the species we see today.

Website: Following is a website providing one model of the Tree of Life theory: https://www.discoverwildlife.com/animal-facts/tree-of-life-evolution

## PRIMORDIAL SOUP THEORY: NOT THE ORIGIN OF LIFE

You may have heard the theory that the early Earth had a "primordial soup" of inert chemicals necessary to produce living organisms. An initial theory was that lightning struck this soup to produce various amino acids, which are the "building blocks" of life. Once the essential amino acids were formed in the soup mix, they proposed life of some kind would evolve over a very long period. For scientists to conduct experiments to produce an amino acid from the soup, they needed to guess what the atmosphere around Earth needed to be like. They assumed it would need to be a hydrogen-rich mixture of methane, ammonia, and water vapor. However, NASA scientists in the 1980s determined that the Earth's atmosphere would have been carbon dioxide, nitrogen, and water. *Science proves that life on planet Earth did not evolve from a primordial soup.*

> Worldview: The theory that a primordial soup was the origin of life has been proven to be untrue by scientists today. However, if it were true, thoughtful people would ask where the Earth and those initial chemicals originated? Answer: If they existed, God created them.

## Original Chemicals Organized Themselves Theory

Another theory by some evolutionary scientists proposes that the original chemicals on Earth organized themselves into complex molecules. In other words, they say non-living chemicals can self-replicate to form living matter. To try this out, go to your kitchen and pull out some household chemicals (baking soda, dish soap, etc.). Pour them together into a bowl and set it aside. Now, wait until the chemicals combine themselves into a living cell. How long do you think you will have to wait? The answer is forever because it will not happen. Something nonliving cannot by itself become living.

Some say if some of these chemicals are catalysts (enzymes), they can help trigger chemical reactions that bring other chemicals to life. This is also not possible since, for it to occur, many chemical reactions must occur in the correct order, in the proper place within the chemical soup, and in the right degree of contribution to the whole. Chemists and ordinary people use catalysts to alter a substance in many ways. Every day, people use baking soda as a catalyst to make bread that will rise.

Instructional Comment: The idea of a primordial soup that was the origin of all complex life on Earth is preposterous. However, if it had existed, the chemicals in the soup would have been created by Jesus out of nothing (*Ex Nihilo*). We know, however, that Jesus did not create life billions of years ago from a bunch of lifeless chemicals.

## CHAPTER QUESTIONS

DQ1: What is the difference between the Scientific Theory of Intelligent Design and the Father God as the biblical Intelligent Designer?

DQ2: How are Evolutionary Creationism and Theistic Evolution the same and different?

DQ3: Describe evolution's biological Tree of Life theory.

DQ4: If this primordial soup had (in theory) existed, where did its chemicals come from?

## DIG DEEP

DD1: Why do evolutionists need a theory like biology's Tree of Life to support their theory of biological evolution?

## PERSONAL APPLICATION

PA1: Why does the existence of so many theories of biological evolution strengthen your faith in the God of the Bible and the creation account?

# PART H
Delving Deeper into Science

# 28.

## Delving Deeper: Archaeology Validates the Bible

Important: Delving Deeper chapters contain material that may be difficult to understand. I suggest you spend extra time studying it and ask for the Holy Spirit's help. The Father and Jesus sent him to live within Christians as their Teacher, Guide, and Helper.

### ARCHAEOLOGY IS IMPORTANT TO SCIENCE AND THE BIBLE

Archaeologists study the material evidence of ancient and recent human cultures. There are three main types of archaeology: (1) Prehistoric archaeology: Study of human cultures that did not have writing; (2) Protohistoric archaeology: Study of human cultures that have incomplete records; (3) Historic archaeology: Study of human cultures that have well-developed historical records. Historical records could be written or oral. Artifacts (tools, weapons, pottery, and such) from these three types provide archaeologists with information about the people, their society, religious beliefs, and culture.

### ARCHAEOLOGICAL STUDY BIBLE

The *Archaeological Study Bible, An Illustrated Walk Through Biblical History And Culture,* is a New International Version (NIV) study Bible that includes archaeological discoveries related to the Bible. It has the typical study Bible features. It offers over 500 short articles on biblical archaeological discoveries, ancient peoples, ancient texts and artifacts, and more. Many articles include full-color photographs. This unique study Bible helps the curious individual explore how the science of archaeology continues to prove the accuracy of the Bible.

### Biblical Archaeological Review

Many archaeological discoveries over centuries validate the Bible. They are referred to as "extra-biblical" evidence. *Biblical Archaeological Review* is a magazine with a newsletter that publishes such extra-biblical archaeological discoveries. Christian scholars and archaeologists write some articles, while others are written by biblical scholars who are not Christians.

Note: A person's attitude about the Bible being the word of God will affect how they interpret the results of archaeological discoveries. For some, their faith may rest in the science they use instead of the God who created what they study.

Instructional Comment: I am not aware of any valid archaeological discoveries that disproved anything in the Bible.

### Unearthing the Bible

In his book *Unearthing the Bible,* 101 *Archaeological Discoveries that Bring the Bible to Life,* Titus Kennedy describes 101 archaeological discoveries that provide compelling evidence of the validity of the Bible. These come from 50 museums, private collections, and archaeological sites. He categorizes and lists them in eight topical chapters. He also describes several extra-biblical myths about creation, including the Babylonian Creation Myth (Enuma Elish) I mentioned earlier. He also vividly describes the creation myths of Girsu (Tell Telloh, Iraq, 2,900 BC); Ebla (Tell Mardikh, Syria, 2,400 to 2,000 BC); Enki and Ninhursag (Nineveh, Iraq, 2,900 to 2,800 BC); and Adapa (Nineveh, Iraq, fourteenth century BC).

The Girsu creation myth is interesting. Its clay tablet, written in Sumerian, tells about the beginning of creation when daylight and moonlight did not shine because the sun and moon did not yet exist. The myth says the land was dust, vegetation had not been created, and the Earth was filled with water.

## ARCHAEOLOGICAL DISCOVERIES THAT VALIDATE THE BIBLE

This section includes archaeological discoveries that have validated the accuracy of the Bible. For a starter, non-Christian archaeologists have contended for decades that the biblical account of Abraham was a myth. They said there was no archaeological evidence to prove its validity. The Bible states that Abram (Abraham) was from the Chaldean city of UR. (This area is modern-day Iraq.) No archaeological evidence had yet proved that UR existed. However, an archaeological dig in Iraq eventually found stone tablets with cuneiform writing that mentioned the city of UR. The very science that wanted to disprove it had validated Abraham's biblical story.

> There are thousands of archaeological discoveries in multiple countries that prove the validity of the Bible. How many prove the validity of the theory of evolution? Answer: None. A theory is not proof.

## Kingdom of David and Solomon Exhibition

The artifacts for this exhibition were discovered at archaeological sites across Israel, including Jerusalem, Timna, Lachish, and Khirbet-Qeiyafa. The 50 artifacts include iron, pottery, stone, and textiles from the tenth century BC. Its Grand Opening in February 2024 provided additional proof of the integrity of the Bible for one of its most renowned figures, King David, and his son and successor, King Solomon. It is hosted by the Armstrong Institute of Biblical Archaeology and the Armstrong International Cultural Foundation in Edmond, Oklahoma.

## Additional Archaeological Discoveries Confirming the Bible

The following archaeological discoveries confirm the accuracy of the Bible. Some of these are included in Titus Kennedy's book.

> Tel Dan Stone: The letters "BYTDWD" on the stone refer to the "House of David." The stone is dated to the ninth century BC and refers to the lineage of King David in Israel. The Mesha Stele/Moabite Rock had the same letters "BYTDWD." It also dates back to the ninth century BC and solidifies the existence of King David of Israel.
>
> Ketef Hinnom Scrolls: The oldest surviving texts from the Hebrew Bible. They include a priestly blessing dated to 600 BC and text from the Old Testament Book of Numbers.
>
> Pontius Pilate's Grand Avenue: Pilate was the Roman Prefect who ordered the crucifixion of Jesus Christ about 33 AD. Archaeologists uncovered a 2,000-foot-long boulevard in Jerusalem built by Pilate. This proves the existence of Pontius Pilate during the time of Jesus.
>
> Dead Sea Scrolls: Discovered in 1947, these were considered one of the greatest archaeological finds ever. They were discovered in caves on the northwestern shore of the Dead Sea. The scrolls included the entire Old Testament except for the book of Esther (38 of 39 books). They were written around the late third century BC before the birth of Jesus Christ.
>
> Mernaptah Stele: This is an engraved stone slab ordered by Pharoah Mernaphat to record his victories over his enemies. It states, "Y-s-rl, his seed is scattered," referring to the nation of Israel. This is considered to be one of the earliest archaeological records showing the existence of the nation Israel. The Pharoah reigned from 1,213 to 1,203 BC.
>
> Hezekiah's Monumental Inscription: The eighth-century BC inscription was discovered in a refuse heap near a pool connected to Hezekiah's Tunnel. Translated,

it read, "Hezekiah's Pool." This is a confirmation of Hezekiah's existence, another biblical figure.

Uzziah, King of Judah Seal Inscriptions: In 2 Kings 14:21, Uzziah is called Azariah. He ruled over Judah from about 792 to 740 BC. His name appears on two seals of unknown origin. One seal reads, "Belonging to Abiah Servant of Uzziah."

Shema Seal (Jeroboam II, King of Israel Inscription): King Jeroboam II's reign over Jerusalem from about 793 to 753 BC is mentioned in 2 Kings 14:2-29. The seal says, "Belonging to Shema, Servant of Jeroboam."

Ancient City of Shiloh: Its location was first discovered in the fourteenth century AD by Rabbi Ishtori Haparchi. The ancient city site has had multiple archaeological digs. An ancient Jewish olive press and several wine presses dating to the first century AD have been discovered in recent years. Shiloh is mentioned in several verses in the Old Testament, including Joshua 18:1 and Judges 21:19. Shiloh was the first political and spiritual center for the nation of Israel. It was where the Tabernacle was located for 369 years after Israel's 40 years of wandering in the desert (after escaping from Egypt).

### Bible Lands Museum, Jerusalem

This museum contains thousands of ancient artifacts about the people and places of the Bible. Quotes from the Bible are located throughout the museum to place the artifacts in their biblical context. Following are three of these discoveries:

Quadrilingual Darius I Jar: Darius I ("Darius the Great") of Persia is a prominent figure in the Old Testament. He is mentioned in the books of Haggai, Zechariah, Malachi, Ezra, Nehemiah, and Daniel. A Persian calcite jar is displayed with inscriptions in four languages that praise Darius. He lived from 550 to 486 BC and reigned from 522 to 486 BC. He was responsible for letting the people of Israel return from their 70-year exile in Babylon to rebuild their Temple in Jerusalem and restore their Jewish religious life.

Jonah Sarcophagus: This fourth-century AD sarcophagus has three scenes from the Old Testament book of Jonah: (1) One scene depicts Jonah being cast overboard from the ship into the mouth of the giant fish; (2) In another, he is on the shore; (3) In the third, he is covered and protected by the plant to teach him compassion.

Christogram Sarcophagus: This belongs to a Christian woman named Julia Latronilla, who died approximately 330 AD. It depicts several Old and New Testament biblical scenes: (1) Abraham's near sacrifice of his son Isaac (Genesis 22); (2) Miracle in Cana where Jesus turned water into wine (John 2:1-11); (3) Jesus' triumphant entry into Jerusalem before his later crucifixion (Matthew 21:1-11). The circle in the center is one of the earliest Christograms discovered. It's a symbol with the first two letters (*chi, rho*) of the Greek name of Christ.

## Tower of Babel

Dr. Andrew George, a professor of Babylonia at the University of London, says that a baked clay tablet from Babylon (modern-day Iraq) provides evidence the Tower of Babel was real. The tablet depicts the tower (ziggurat) with its seven steps, the king with his conical hat and staff, and describes the commissioning of the tower's construction.

## Sayings of Jesus

This article is in the Spring 2024 Biblical Archaeological Review Magazine. A small papyrus fragment may have some of the earliest recordings of statements by Jesus while on Earth. They revolve around leaving behind worldly cares. These are similar to those recorded in the Gospels of Matthew and Luke. This early second-century AD fragment was discovered at the Oxyrhynchus site in Upper Egypt. This is another archaeological discovery that confirms the Bible's authenticity and the reality of Jesus Christ.

# ARCHAEOLOGY AND THE LIFE OF JESUS CHRIST

In his book, *Excavating the Evidence for Jesus: The Archaeology and History of Christ and the Gospels*, Dr. Titus Kennedy answers such questions as, "How can you know that Jesus Christ was a real person (not a myth) who lived on Earth some two thousand years ago?" "Is there verifiable evidence that He existed?" Dr. Kennedy is a Christian and professional archaeologist who reviews archaeological and biblical evidence of the historicity of Jesus Christ. He is an example of a person of science who is a Christian who writes about his Savior and Lord with accurate biblical and scientific facts. He uses extensive archaeological discoveries in Israel to trace the birth, life, ministry, death, and resurrection of Jesus Christ to verify his existence and life. He follows Jesus through the New Testament Gospels, identifying scientific findings and historical literature to affirm the biblical account of Jesus Christ.

Note: This book is well worth your time to learn about scientific evidence for the life of the Savior and Lord who created and sustains all things. Your faith in the Bible and the creation account will be strengthened as you learn more about Jesus Christ.

## CHAPTER QUESTIONS

DQ1: Describe what archaeology studies and does not study.

DQ2: List four archaeological discoveries you knew about before reading this book that proved the Bible's accuracy.

DQ3: Why haven't non-biased reports of archaeological discoveries disproved anything in the Bible?

DQ4: Describe why the Dead Sea Scrolls are considered one of the most important archaeological discoveries ever.

## DIG DEEP

DD1: Why is the unbiased study of archaeology an important element in validating the reality of the God of the Bible?

## PERSONAL APPLICATION

PA1: How do these archaeological discoveries help you believe the Bible is God's inspired word (2 Timothy 3:16)?

# 29.

# Delving Deeper: Sciences Validate Creation

Important: Delving Deeper chapters contain material that may be difficult to understand. I suggest you spend extra time studying it and ask for the Holy Spirit's help. The Father and Jesus sent him to live within Christians as their Teacher, Guide, and Helper.

## ATHEISM, NOT GOD, IS DEAD

A Time magazine article, "Is God Dead?" in 1966, brought atheism to the forefront of American thought. It became trendy to believe there was no God. However, since then, the idea that God does not exist has become less popular. Eric Metaxas's book, *Is Atheism Dead?* is based on observations that atheism is far less acceptable than before. Today, we see that the sciences validate the creation story and, therefore, the existence of God. Eric believes that many people today see atheism as irrational. So, he says that atheism is dead, not God.

*Is Atheism Dead?* addresses much more than atheism. It brings together science and the Bible to explain the created universe. Eric addresses modern archaeological advancements, historical evidence, and testimonies from scientists in his detailed analyses. His thirty topical chapters clearly and logically engage readers in the wonder of God as the Intelligent Designer of everything and the universe in particular. His topics include the "Big Bang Theory," the origin of life, the failure of evolution to explain our world, and much more. He sees the universe and our planet Earth as being "fine-tuned" by Jesus. For example, this fine-tuning enables Earth to be inhabitable for all living organisms, particularly people. He says that more and more modern scientific discoveries validate creation, God as its Designer, and Jesus as its Sustainer (Colossians 1:16–17).

## CREATION AND EVOLUTION: MANY SCIENTIFIC DISCIPLINES

Following is a list of some of the sciences that can be involved in the study of creation and evolution. Some scientists use their scientific knowledge to attempt to prove the theory of the evolution of species and the evolution of the universe. In contrast, those scientists who are creationists strive to validate the accuracy of the Bible and the creation account.

> Scientists who are creationists use the sciences to continually validate the accuracy of the Bible and the creation account.

Any of the following sciences can be used for either purpose by scientists:

- Anthropology: Study of what makes people "human." In other words, it studies the origin and development of human societies and cultures, including languages, beliefs, morality, institutions, and material goods.
- Archeoastronomy: Study that combines archaeology, astronomy, and anthropology to study how prehistoric and ancient people used the stars and constellations for seasons, solstices, navigation, and more.
- Archaeology: Study of graves, buildings, tools, and other objects of ancient and modern people. Archaeologists do not directly study human remains. Rather, they study how they lived, worked, hunted, and existed.
- Astrophysics: This branch of astronomy studies the physical nature of stars, planets, and other phenomena in the universe. These scientists apply the laws of physics and chemistry to understand how the universe works. They use massive telescopes and data collected from satellites and space probes. They may study such phenomena as the birth, life, and death of stars, galaxies, black holes, and other physical objects in outer space.
- Biology: Study of life, both living and ancient. Biologists research how organisms form, develop, and interact with each other and their environment. Because of this, some consider biology to be the cornerstone of evolution.
- Genetics: Study of genetic material and information, including heredity. It studies genes and everything related to them, including the human genome, chromosomes, DNA, and so on. Geneticists can be involved in studying ancient skeletons and other ancient matter.
- Morphology: Study of the form and structure of organisms.
- Paleoanthropology: Study of ancient humans and creatures similar to humans (such as apes).
- Other sciences may also be used to study creation and evolution, including mathematics, physics, and chemistry.

## CHRISTIANITY AND SCIENCE ARE COMPATIBLE

Some people think science and Christianity are polar opposites. They think you can't believe in both but must choose one over the other. Central to this conflict is the theory of biological evolution and the opposing Christian belief of creation by an Intelligent Designer. Some people fail to understand that evolution is still just a theory that science can't prove. However, science has proven the validity of the Bible and Christianity.

Following is a true story about a discussion about this I had with a young Christian:

> Some years ago, I discussed science and Christianity with a young Christian lady in an adult Sunday School class I was teaching. She mistakenly thought evolution was proven science. Due to her strong belief in creation, she considered all science to be opposed to the Bible and Christianity. I told her evolution was a theory and not a proven science. I said that actual science and the Bible are not only compatible but inseparable because unbiased science is a study of God's creation. I explained that the Bible was replete with verses related to the sciences, such as biology, physics, astronomy, geography, and more.

Important: After all, since the Triune God created everything, why shouldn't the study of science, without bias, be considered important study of the natural world he created?

## BIBLE AND SCIENCE

Both the Bible and science reveal the wonders of God and his creation. Many Bible verses support this complementary relationship. This corroboration is further evidence that validates the existence of an Intelligent Designer. From their scientific discoveries and analyses, more scientists confirm what God says about creation in the Bible.

Following are a few scientific disciplines and related facts about the natural world that are found in the Bible:

- Geology: Fountains (springs) of water exist under the seas (Genesis 7:11); The flow of wind currents and their direction (Ecclesiastes 1:6)
- Physics: The Earth is round (Isaiah 40:22); Gravity holds the Earth up in space and its stars and planets in their places (Job 26:7)
- Hematology: The life of people and all animals is in their blood (Leviticus 17:11)
- Oceanography: The flow of warm and cold ocean currents (Psalm 8:8)

Unfortunately, some scientists don't want to admit that when they spend their lives studying nature, they are also learning about the Creator God. Some pursue their own agenda and bias.

## DISCUSSION QUESTIONS

DQ1: What is the difference between atheism and agnosticism?

DQ2: Which of these scientific disciplines could scientists who are Christians use to prove the validity of the Bible?

DQ3: Find three additional verses in the Bible that reflect the sciences.

DQ4: If you could pick only one science that validates creation and disproves evolution, which would it be, and why?

## DIG DEEP

DD1: What would you say to someone to help them understand that the Bible and science (without bias) do not contradict one another?

## PERSONAL APPLICATION

PA1: How would you respond to a close friend who told you that God was dead and that he did not exist?

PA2: Which of the sciences do you feel most closely aligns with your thinking about the natural world created by God? Why do you feel this way?

# 30.

# Delving Deeper: Intricate Design Validates Creation

Important: Delving Deeper chapters contain material that may be difficult to understand. I suggest you spend extra time studying it and ask for the Holy Spirit's help. The Father and Jesus sent him to live within Christians as their Teacher, Guide, and Helper.

## INTRICATE COMPLEXITY BY GENETIC DESIGN

The minute and complex intricacies of the design of plants and animals, especially humans, point to the need for an Intelligent Designer. The Father God designed everything using intricate and complex designs only he could imagine. From his unfathomable imagination, he created genetic information that drives everything, including traits, behaviors, and functions.

> Every part of the human anatomy and physiology demonstrates God's intended design for function and purpose.

The following verses tell us that people are God's marvelous workmanship:

> You made all the delicate, inner parts of my body and knit me together in my mother's womb. Thank You for making me so wonderfully complex! Your workmanship is marvelous—how well I know it. You watched me as I was being formed in utter seclusion, as I was woven together in the dark of the womb. (Psalm 139:13–15 NASB)

Bible Commentator Albert Barnes[1] describes these verses as follows:

> The reference here is to the various and complicated tissues of the human frame—tendons, nerves, veins, arteries, muscles, "as if" they had been woven, or as they

---

1. Barnes, *Notes*, Psalm 139:13–15.

appear to be curiously interweaved . . . . no art of man could "weave" together such a variety of most tender and delicate fibres and tissues as those which go to make up the human frame . . . and who but God could "make" them? (Albert Barnes' Notes on the Bible)

Let's look at two examples of intricate complexities that support the biblical creation account.

## Human Brain

*The human brain is the most complex organ in the human body and the most complex entity in the known universe.* It contains about 86 billion neurons and 85 billion other cells. The neurons communicate with each other through over 100 trillion connections to regulate various functions and processes of the human body. How could billions of intricately connected cells evolve from a one-cell organism to become the human brain we have today? Consider that your brain produces your thoughts, actions, feelings, memories, and sensory experiences with the world around you. Does the evolution of this highly complex functional organ seem likely to you?

Website: More about the staggering complexity of the nature and functioning of the human brain can be read on the following website: https://www.psychologytoday.com/us/blog/consciousness-and-beyond/202309/the-staggering-complexity-of-the-human-brain

## Human Eye

The human eye is another intricate and complex organ in the human body. It consists of about 40 individual subsystems, including the retina, pupil, iris, cornea, lens, and optic nerve. It is estimated that it can process 36,000 bits of information in an hour. It adjusts to head movements within milliseconds. The retina has about 137 million special cells that respond to light and send messages to the brain. It has about 130 million light-sensitive rods and cones that convert light into chemical impulses. The brain's visual cortex interprets and converts these impulses to color, contrast, depth, etc., allowing you to see "pictures" of the world in which you live. Does it seem likely that the intricate and complex human eye could have evolved from a one-celled organism?

Website: More about the amazing complexity of the nature and functioning of the human eye can be read on the following website: https://www.allaboutthejourney.org/human-eye.htm

Important: The intricate complexity of human beings is evidence that it is a product of divine engineering by the Father God, the Intelligent Designer.

## CREATIVITY, INNOVATION, AND INVENTIONS

The Father God designed the human brain with incredible abilities that enable people to go beyond the depths of human reasoning. One of these sentient abilities is called "creativity." When applied, it can result in innovations and inventions that can dramatically improve the quality of human life.

Definition: *Creativity* is the capacity or trait to conceive of something not previously known to the individual or society in general.

*Innovation* is the application of a creative idea that results in a new (not necessarily unique) product, service, or way of doing something. It can also significantly improve something that already exists.

*Invention* results from an innovative idea that produces a unique product, service, or process (not just an improved one). Not all innovative ideas lead to unique inventions.

## SENTIENT NATURE OF THE HUMAN SOUL

Being sentient means people can feel and sense things within and around them. It's self-awareness. It means they have a soul that is beyond the physical world. Evolution can only attempt to explain the development of the natural and physical. So, how did the soul of people (and animals) come into existence? There is only one way: the Creator God designed and created it.

The sentient nature of the human soul includes God-designed capacities and characteristics as follows:

- Consciousness (self-awareness)
- Emotions/Feelings
- Intelligence (capacity to learn, adapt, and use knowledge)
- Reasoning/Thinking
- Will (Self-determination, inclinations, and decision-making; self-will)
- Imagination (capacity to envision what does not exist)
- Memories (mentally stored experiences that can be used later)
- Creativity (leads to innovation and invention)
- Conscience/Morality

### Sentient Nature in Animals?

Scientists differ in opinion about whether animals have a soul and are sentient. Part of the issue concerns whether they are self-aware and have a conscience. Some don't think animals are creative or have an imagination. If you have a pet, especially a dog or cat, you can testify that they can reason and think, have emotions and feelings, and have memories. So, I believe they have a soul, but not one exactly like the human soul.

### Non-Sentient

When creatures respond to stimuli without involving thinking or cognitive processing, they are not being sentient by their nature. They respond only by innate stimulus and response mechanisms. For example, bacteria are not sentient. Artificially intelligent creations (AI-enabled) and robots are not sentient; they are not self-conscious beings with a soul.

## HUMAN MORALITY VALIDATES CREATION

Morality exists only within a social system. It is derived from what people determine to be right and wrong. For example, in Western cultures, helping someone in need is right. However, murdering that same person is wrong. Societies create a judicial system to manage and hold people accountable to their instituted moral codes. Since the Creator God is a moral being, he designed people with the ability to determine right and wrong through the conscience. He does so to hold people accountable for following his moral codes of behavior clearly defined in the Bible and guided by the Holy Spirit. A society without just and righteous moral codes dictated by a properly developed biblical conscience will eventually devolve into chaos.

### The Human Conscience

People have a conscience as part of the image of God (it's part of the human soul). It's a warning system from God concerning violations of a moral code of conduct. It's an inner sense (self-awareness) that judges a person's actions, words, and motivations.

Following is some additional information about the human conscience:

- The conscience is a God-designed capacity that exists at physical birth (its capacity is like a blank place that needs to be filled in).
- Its substance (moral codes) is developed over time as the brain and mind mature organically. (If a newborn baby dies, it does so without utilizing the capacity of a conscience.)
- Moral codes of conduct develop in two ways: (1) They are unconsciously absorbed from culture, society, and teachings; (2) They are intentionally learned as people consciously engage with their surroundings, culture, society, and relationships.
- People can choose to change the moral codes of their conscience, that is, what they accept as right and wrong.

- Habitual sin corrupts a non-Christian's mind and conscience (Titus 1:15) and defiles a new Christian's weak conscience (1 Corinthians 8:7). The conscience can mislead a person if it's faulty.
- The sacrificial death of Jesus Christ brings forgiveness of all sins and cleanses a new believer's conscience of all rightful guilt (sin). Therefore, there is never condemnation, never a rightfully guilty conscience from God about forgiven past sins (Hebrews 9:14).
- The indwelling Holy Spirit can guide the Christian's conscience (Romans 9:1).

Note: people growing up in different cultures will likely develop different moral codes of conduct. These ideas of right and wrong can be changed as they adapt to different cultures, societies, teachings, and so on.

## CHAPTER QUESTIONS

DQ1: Why do the minute and complex intricacies seen in the design of plants and animals point to the need for an Intelligent Designer?

DQ2: What would this world be like if God had not designed the human brain with creativity, innovation, and invention?

DQ3: How would someone who believes in biological evolution explain the soul's existence and sentient nature?

DQ4: How do people develop their moral codes of conduct?

## DIG DEEP

DD1: People in different cultures can have different codes of conduct. Why does this indicate that God created people with the ability to develop a code of conduct but not with a specific code?

## PERSONAL APPLICATION

PA1: When was the last time you had a creative idea, perhaps even a simple one that improved your daily routine, for which you thanked your creative God?

PA2: Based on the descriptions of the human conscience, why do you think the statement, "Let your conscience be your guide," is not the best option for you?

# 31.

# Delving Deeper: Genetics Validates Creation

Important: Delving Deeper chapters contain material that may be difficult to understand. I suggest you spend extra time studying it and ask for the Holy Spirit's help. The Father and Jesus sent him to live within Christians as their Teacher, Guide, and Helper.

## GENETICS IS COMPLEX

The modern science of genetics is broad and complex, so it is not realistic for me to describe it in detail. Instead, I will focus on how it relates to creation and the theory of biological evolution.

Definition: *Genetics* studies an organism's chromosomes, genes, DNA, RNA, and other genetic factors to determine traits and how and why the organism functions. Genetic information is used in many sciences to discover why an organism looks and behaves as it does.

### Many Uses for Genetics

Genetics is used in many ways to benefit people. For example, it is used in medical science for diagnosing and treating diseases, identifying hereditary predispositions and defects, vaccinations, and in non-medical areas such as family ancestry research.

## CHROMOSOMES, GENES, AND DNA BASICS

Every person has 46 chromosomes and about 20,000 genes located within nearly every cell of the body. Half of the genes are inherited from the mother and half from the father on each parent's 23 sets of chromosomes. Genes are composed of DNA molecules, which have genetic information encoded in the DNA. The human genome comprises all chromosomes, genes, and DNA of a human being. Genes and DNA dictate every aspect of an individual.

---
Genetic composition and functioning in people reveal organized and designed complexity. How is it possible that the human genome evolved by random chance?
---

Definition: *Allele*: One of two or more variations of a gene on the same place on a chromosome. Each parent provides one version of the allele. Different alleles can result in different traits.

*Chromosome*: Thread-like strings of DNA tightly coiled around proteins inside the nucleus of cells. People have two pairs of 23 chromosomes, one pair from each parent, for a total of 46.

*DNA*: Most Deoxyribonucleic acid (DNA) is located in the nucleus of cells, while some are also found in the mitochondria (known as mtDNA). DNA is passed on to offspring. The mtDNA in cells converts the energy from food into forms the cells use.

*DNA letters*: DNA stores its genetic information in a string of four chemicals: guanine (G), cytosine (C), adenine (A), and thymine (T). Geneticists use the letters to analyze genetic information and in genome projects to study the sequence of the DNA letters.

*Gene*: Basic unit of genetic information that occupies a fixed position on a chromosome. It's composed of DNA.

*Gene pool*: Collection of all the different genes within the population of a species.

*Heredity*: Passing of traits (from genes) from parent to child.

*Human genome*: All the genetic information for human beings.

*Mutation*: Change in the gene's biological structure. A change in the DNA sequence of a gene.

*Trait*: Physical or behavioral characteristics from genes (passed from parent to child).

## Chromosomes and Gender

People have 23 pairs of chromosomes from each parent, for a total of 46 chromosomes embedded in their cells. The various genes on the chromosomes dictate physical traits and behaviors. In other words, they determine the form and function of all living creations.

The human sex chromosomes of the female egg (XX) and male sperm (XX or XY) determine a child's gender. The male who contributes the XY chromosome (along with the female XX) causes the child to be male. The male's XX and female's XX chromosomes result in a female child.

Website: More information about chromosomes, genes, and reproduction can be found on the following website: https://www.genome.gov/about-genomics/fact-sheets/Chromosomes-Fact-Sheet

## Jesus Christ Incarnate Chromosomes and Genes

The human genome is all of the genetic information needed for human life. This means every human being has 46 chromosomes with all their genes. As you know, Jesus Christ was born to his human mother, Mary, by the Holy Spirit. He did not have a human father. Since the Holy Spirit is a spirit and does not have a physical body, he did not physically possess 23 human chromosomes with genes to pass on to Jesus at his conception. So where did the 23 chromosomes with the Male XY chromosomes come from?

Instructional Comment: We know that Jesus was fully human (Mother's 23 human chromosomes) and fully God (Holy Spirit) at conception. For Jesus to exist as a human being with 46 chromosomes, perhaps the Holy Spirit supernaturally created 23 human chromosomes, including the XY chromosome. These would have merged with Mary's egg containing her 23 chromosomes at the conception of Jesus.

## Resurrection of Jesus Christ and Chromosomes

The human, physical body of Jesus died and was buried. Therefore, his human cells, where chromosomes and genes reside, died because no blood and oxygen can circulate after death. When the physical body of Jesus was resurrected, all his cells were given eternal life. This means all his chromosomes and genes were no longer human but were eternal. Jesus now sits on his throne in Heaven with his eternal body and an eternal genome.

But what about the resurrected bodies of the followers of Jesus Christ? Will they also have eternal cells and chromosomes, an eternal genome? Perhaps they will since they will be eternal beings like Jesus.

## Do Angels Have Chromosomes and Genes?

Angels are created eternal spiritual beings. They are ephemeral yet can possess some physical form. For example, they stand before the throne of God in Heaven, come to Earth to serve people and speak with people (for example, Gabriel with Mary). Cells (with their chromosomes) are the substance of every living physical creation. Without cells, there is no physical life. Perhaps Jesus created eternal angels with eternal cells and chromosomes different from human cells and chromosomes that perish. Since they were not made from the 92 natural elements as people were, perhaps they were made from spiritual elements.

## Adam and Eve's Chromosomes and Genes

Jesus created Adam and Eve as perfect eternal beings, so they must have had eternal cells and chromosomes. Only after the Fall did their eternal body become corrupted and human as ours are today. In other words, only after their rebellion against God were their cells and chromosomes able to die. But how many chromosomes did they each have when they were created? Did the Father design and Jesus create them to have 46 chromosomes since they didn't have parents to inherit 23 pairs of chromosomes from? Adam and Eve were created to reproduce so they each could give 23 sets of chromosomes to their offspring.

## GENETICS FROM A CREATIONIST'S PERSPECTIVE

The study of genetics by creationists has provided information about how the Intelligent Designer God used genetic information to create everything living. It also provides insights into how all living things can adapt and change to survive. (This is microevolution, not macroevolution.)

## ENVIRONMENT, RECOMBINATION, AND MUTATION VALIDATE CREATION

Three concepts in genetics help validate creation and disprove the Theory of Evolution by Natural Selection. These are environment, recombination of genes, and mutation of genes. This may seem complex at first, but if you read carefully, you may find they are not difficult.

### Environment

Evolutionists say that an organism's physical traits can be altered through life-long environmental exposures. For example, some say that life-long exposure to the sun can darken a person's skin color, which will be passed on to offspring. However, these environmental-induced changes cease at death and do not affect genes. Therefore, they are not passed on to offspring. An organism's lifetime environmental exposures are not natural selection and adaptation and do not result in microevolution or macroevolution.

### Recombination

The recombination of genes is interesting. Mendel discovered this. It's a way of saying that organisms don't use all their available genes in one generation of the species. Genetics shows that each subsequent generation usually has all the genes that existed in the original generation (both parents). Each subsequent generation retains all the same genes but only uses some of them. For example, biological children carry both parents' genes but can have different traits from different genes. The next generation, produced by these children, can also have different traits also because of different genes.

### Mutation

The mutation of existing genes can cause a different trait to appear. They are mistakes in the genetic copying process of DNA. Cells usually correct these errors. Occasionally, an incorrect (mutated) gene is not corrected and becomes part of the organism's genetic information. These randomly created mutated genes that benefit the organism may enable it to survive, while harmful ones may result in it not surviving. This is part of the process of natural selection guided by genetic information by which an organism adapts or fails to adapt to a new or changing natural environment (microevolution).

Website: Detailed information about genetics, how it disproves the theory of evolution, and how it validates creation can be found on the following creation website: https://creation.com/genetics-no-friend-of-evolution

Important: Evolutionists incorrectly theorize that sufficient beneficial gene mutations can cause a species to evolve into a new species or genus (macroevolution). This has never been proven to be true.

## HUMAN GENOME PROJECTS PROVE ONE HUMAN SPECIES

Evolutionary geneticists and other scientists have been trying to understand the human genome. To do so, they have been studying DNA sequencing to determine the exact sequence of DNA nucleotides (bases) in a DNA molecule. They use the first letters of their chemical names in sequencing: guanine (G), cytosine (C), adenine (A), and thymine (T). Each unique sequence provides biological information that tells a cell how to develop and function. They are keys to how individual genes function.

The following are two human genome projects that have provided significant information on human genetics. *The results point to one biological, genetic human family to which all people are related.* Evolutionists say these studies prove that modern humans evolved over a long period from ancient non-human ancestors. *Creationists know that Adam and Eve were the first humans to which all people are genetically and biologically related.*

Instructional Comment: Creationists appreciate the massive scientific efforts to identify the biological roots of humanity. We, however, know that the first modern humans were Adam and Eve, who did not live in Africa. The Bible tells us about the Garden of Eden, where they lived before their fall into rebellion and sin against their Creator God. (Genesis 2:8–14). This area is modern southwestern Asia, not Africa.

Following are brief descriptions of the two studies.

> Study #1 (2003): One group of global scientists provided their initial human genome results in 2003. It was considered one of the most ambitious projects ever developed, including comparisons to such massive projects as splitting the atom and going to the moon. Their goal was to sequence the three billion DNA letters (G, C, A, T) in the human genome. All the DNA of a living organism is called a genome. The completed project sequenced about 99 percent of the human genome with an accuracy of 99.99 percent. This allowed scientists to develop a better understanding of the genetic makeup of human beings. One conclusion of the study was that all humans biologically and genetically belong to one human race (one human species).

Website: The following website provides details of the project and its results: https://www.genome.gov/human-genome-project/results

> Study #2 (2022): Another more recent global team of scientists combined their genetic study data of 3,609 individual genome sequences from 215 people groups

to produce a massive evolutionary family tree. They theorize that the tree identifies nearly 26 million ancestors of modern humans and where they lived globally. This research, they say, proves that modern humans evolved from non-human ancestors originating in the Sudan, Africa, who lived there about one million years ago. They believe this new genealogical mapping technique proves the theory of human evolution. They state this genealogy tree demonstrates that all modern people are related and come from one ancient genetic source. However, as creationists, we know this study and its "family tree" validate the creation account that Adam and Eve were the genetic and biological parents of all people.

**Animal and Plant Genome Projects**

The scientists who studied the human genome said this genetic mapping technique can also be used for animal and plant genome studies. However, they said the method is not fully accepted by some within the scientific community.

## INTELLIGENT CELL FUNCTIONS

How does each cell in the body know its purpose and function? The human body contains about 37.2 trillion cells. Since cells contain the same chromosomes, DNA, and genes, how does each cell know what to do? In theory, they each have the same capabilities. So, how does a cell know what kind of cell it is? Is it supposed to be a brain cell, heart cell, leg calf cell, skin cell, and so on? How do cells that are part of the various "autonomic nervous systems" know to never switch off (become inactive)? Are these mysteries beyond human comprehension? No, these are not mysteries because science once again proves that the Intelligent Designer, Father God, designed intelligence into cells to determine their purpose and function.

God, in his ingenious designs of humans (and everything in nature), built into every human cell the ability to know what it is, what to do, and when to do it. Simplifying something very complex, human DNA contains instructions that ensure each cell has all it needs to perform its functions. It's a multifactor, highly regulated process required to provide proper cellular production.

Website: More information about how cells know what to do in the body can be found on the following website: https://www.livescience.com/how-dna-turns-on-off.html

## RANDOM GENE MUTATIONS

The classical description of natural selection is that it's a process by which species (plants and animals) better suited to their environment adapt to survive and reproduce successfully due to beneficial random, non-directed genetic mutations. In this theory, no cellular intelligence directs the changes to genes. They are random and occur by chance. This theory assumes successful adaptation enables a species to survive and evolve into a new species (macroevolution.)

Important: By random and non-directed, I mean these mutated genes are created due to an unplanned mechanism within the cell. However, you just read how random mutations occur at the cellular level of DNA copying. So, yes, because we cannot predict when a cell will make a mistake copying DNA and creating a mutated gene, they can be considered random and non-directed gene mutations.

## INTELLIGENT NON-RANDOM GENE MUTATIONS

Newly discovered non-random gene mutations are understood to occur as cells sense their changing environment and target specific changes to genes that alter the organism's physical and behavioral traits to adapt successfully. The cell intentionally, not accidentally, creates genes with specific traits to enhance environmental adaptation. Since these genes are not an exact copy of another gene, they are mutated genes. In other words, every cell's built-in intelligence tells it what its purpose is and how to enable adaptation for optimum survival.

Important: This is not the DNA copying process where accidental mistakes randomly create mutated genes.

### Random Verses Non-Random Gene Mutation Conference

Some evolutionary biologists met at a conference in Lisbon, Portugal, in 2017. The conference theme was "On the Nature of Variation: Random, Biased and Directional." In other words, their purpose was to validate that some mutations are random and undirected while others are non-random and directed by cells. This thinking was a radical change for mainstream evolutionary biologists who believed all genetic mutations occur randomly by chance without any direction or design. The author of this article, Randy J. Guliuzza, says there are hundreds of technical, scientific papers validating these non-random, cell-directed mutations. However, these evolutionary biologists did not express the belief that the Creator God of the Bible made plants and animals with the ability to change their genetic makeup to adapt to their environment.

Definition: *Random gene mutations* are said to occur randomly, without direction. They are the result of errors within a cell as it copies DNA. Some are beneficial, and some are harmful. Those that are beneficial enable an organism to adapt to its environment successfully.

*Non-random gene mutations* occur as cells sense their changing environment. In response, they create specific changes to genes that alter the organism's physical and behavioral traits to adapt successfully to the changing environment. These mutated genes sensed as harmful to the organism are typically destroyed within the cell. Some escape the cell's efficient monitoring and become part of the organism's genetic composition.

Note: Many modern evolutionists no longer support the theory that random, non-directed mutations occur and drive adaptation and macroevolution.

## INTELLIGENT ENGINEERED GENE CODES

Additional recent studies of genetic code reveal that multi-layered instructions exist within cells. These intricate layers of cellular instructions are part of the cell's ability to manage gene mutations. It's like a series of designed intelligent checks and balances that control cell functions at the genetic level. Intelligent design is programmed into the cells of all living organisms. This is another part of God's design to enhance cellular gene management.

Website: This very technical article written by Jeffrey P. Tomkins and published by ICR can be found on the following website: https://www.icr.org/article/14831/

## CHAPTER QUESTIONS

DQ1: Why do you think every follower of Jesus Christ should have a basic understanding of how the science of genetics validates the Bible and disproves the theory of evolution?

DQ2: Describe genetic recombination and how it affects a human.

DQ3: What is the difference between random and non-random gene mutations?

DQ4: Why does the existence of engineered gene codes require an intelligent designer?

## DIG DEEP

DD1: What do guanine (G), cytosine (C), adenine (A), and thymine (T) have to do with DNA and cellular function?

## PERSONAL APPLICATION

PA1: Trying to understand the genetic component of creation is difficult. Has this short discussion about genetics enhanced your understanding of how it supports creation and disproves evolution?

# 32.

# Delving Deeper: Genetics and Creation Kinds

Important: Delving Deeper chapters contain material that may be difficult to understand. I suggest you spend extra time studying it and ask for the Holy Spirit's help. The Father and Jesus sent him to live within Christians as their Teacher, Guide, and Helper.

## CREATION KINDS AND EVOLUTION OF SPECIES

Genesis 1:20–25 tells us that "kinds" of creatures were made. These kinds were probably a master species (my term). Each creature's significant and diverse genetic composition allowed them to develop and adapt to different locations and environmental conditions over thousands of years (microevolution). The enormous amount of genetic material Jesus placed within each creation kind explains this diversity in plant and animal life. For example, about 118 varieties of domestic dogs, along with wolves, jackals, and coyotes, are linked genetically to the same creation kind. Horses, donkeys, and zebras are the same creation kind. As each biblical kind was fruitful and multiplied, genetic recombination, natural selection, and favorable gene mutations resulted in the great diversities we see today.

> The enormous amount of genetic material Jesus placed within each creation kind explains the diversity in plant, animal, and human life.

This does not mean that creationists believe in the evolution of new species over millions and billions of years. Instead, this refers to the idea that these kinds adapted to their unique environments and developed into biologically related species (and subspecies) over thousands (not millions and billions) of years. Because these related species came from the same creation kind, they carried sufficiently similar genetic material, allowing them to interbreed. This is not the biological evolution of a new species, which is macroevolution.

Definition: *Genus* is the level above species in the biological classification of organisms. It's composed of a select number of species related to one another genetically.

*Species* are groups of organisms related to one another in a specific genus that can reproduce between themselves but not outside their genus. For example, human beings are classified in the genus "Homo" and species "sapiens" (Homo sapiens).

### Creation Kinds and Reproduction

In the following verses, we see Jesus speaking to create creation kinds and commanding them to reproduce and fill the Earth.

> God said, "Let the water swarm with swarms of living creatures and let birds fly above the earth across the expanse of the sky." God created the great sea creatures and every living and moving thing with which the water swarmed, *according to their kinds*, and every winged bird according to its kind. God saw that it was good. God blessed them and said, "Be fruitful and multiply and fill the water in the seas, and let the birds multiply on the earth." There was evening, and there was morning, a fifth day. God said, "Let the land produce living creatures *according to their kinds*: cattle, creeping things, and wild animals, each *according to its kind.*" It was so. God made the wild animals *according to their kinds*, the cattle *according to their kinds*, and all the creatures that creep along the ground *according to their kinds*. God saw that it was good. (Genesis 1:20–25 NET, author's emphasis)

*Kind*, Hebrew is *mîyn* (noun); meaning a sort or kind of something.

The Hebrew *mîyn* is a general type, which is what the translated word kind means. It can also be translated as species, meaning a specific type or kind of something.

## GENE VARIATIONS AND CREATION KINDS

Alleles are variations of a gene in the same place on the chromosome pair from each parent. The different alleles can result in different traits for a person. The common understanding of dominant and recessive genes is an example of different alleles creating different traits. Therefore, alleles are involved in the genetic diversity we observe in organisms since they are the basis of hereditary variations. They influence physical traits, disease susceptibility, and much more. These are examples of the creative diversity the Father designed into human cells and their genetic functions for his creation kinds.

## NO NEW GENETIC MATERIAL FOR EVOLUTION'S NEW SPECIES

Evolutionists believe that new genetic material spontaneously appears within a species, enabling it to evolve into a different species. It's important to know that this has never been known to occur. No new genetic material has been proven to come into existence for a new species. For example, the only way a water-based amphibian can evolve into a land-based reptile is if that amphibian is miraculously given massive amounts of new reptile genetic material. If this did occur (it doesn't), it would be a miracle that only God could perform and would not be the product of natural, biological evolution.

## DIFFICULTY IDENTIFYING SPECIES

Proving the direct line of evolutionary descent of species is not possible. There are no transitional species (missing links) to support this theory. Scientists also find it difficult to identify species due to the significant variation in genetic material and the wide range of species within a genus. Since the original creation of kinds, species have been losing some genetic material while other genes have been corrupted. This, too, has created biological classification problems. Because of modern genetics's complexity and the diversity of opinions about its usefulness, scientists cannot agree upon a universal way to identify and determine new species.

Website: The following evolution-based website provides more information about the difficulties scientists are having in identifying species: https://www.nature.com/scitable/topicpage/why-should-we-care-about-species-4277923/

Instructional Comment: I suspect the loss and corruption of genetic material started from the fall and corruption of Adam and Eve.

Note: About 2.16 million species have been identified and described. The problem is how these species are related to one another.

## CLADOGRAMS

Another way scientists attempt to prove the theory of biological evolution is through the use of cladograms. These don't require proving a one-to-one direct line of evolutionary descent. Many scientists use cladograms to depict a hypothetical relationship between groups of animals visually, how they are related, and their theoretically most common ancient ancestor. Cladograms visually depict particular physical traits of plants and animals to allow scientists to infer an evolutionary relationship to ancient species having those characteristics. For example, one theory is that modern birds with feathers must have evolved from a common ancient ancestor with feathers. Other scientists use DNA sequencing from modern species to compare traits for species with cladograms. The hypothetical relationships in cladograms eliminate the need to theorize a chain of one-to-one evolutionary relationships and missing links.

Note: Cladograms do not describe the process of evolution, species lineage, and the biological Tree of Life. Also, DNA doesn't show how different species are related to each other. It can only show that they are related.

Website: More information about cladograms can be found on the following evolutionary website: https://biologysimple.com/cladogram/

## CHAPTER QUESTIONS

DQ1: Why do you think creation kinds explain the different varieties of plants and animals we see today?

DQ2: Alleles are variations of a specific gene in the same place on the chromosome pair for each parent. Two different alleles can result in a different trait for each child. Why is a recessive gene an example of this?

DQ3: Evolutionists believe that new genetic material spontaneously appears within a species, enabling it to evolve into a different species. Why is this not true?

DQ4: Many evolutionary scientists disagree on a method to identify and classify new species. How does this disagreement undermine the credibility of the biological evolution of the species?

## DIG DEEP

DD1: The author sees each creation kind as a master species that carries an enormous amount of diverse genetic material. Explain why this master species concept might be true.

## PERSONAL APPLICATION

PA1: How would you help a brother or sister understand why they had a different eye and hair color from you?

# PART I

Creation of the Universe

# 33.

# Creation of the Universe

## UNIVERSE WAS CREATED

Genesis 1:1 says, *In the beginning, God created the heavens and the earth* (ESV).

> Praise him, sun and moon, praise him, all you shining stars! Praise him, you highest heavens, and you waters above the heavens! Let them praise the name of the LORD! For he commanded and they were created. And he established them forever and ever; he gave a decree, and it shall not pass away. (Psalm 148:3–6 ESV)

The Bible tells us the universe was spoken into existence by Jesus Christ.

> The LORD *merely spoke, and the heavens were created.* He breathed the word, and all the stars were born. (Psalm 33:6 NLT, author's emphasis)

## HOW MANY STARS AND PLANETS WERE CREATED?

No one knows how many objects exist in the universe. Scientists make their best guesses from observations, mathematical calculations, and assumptions, knowing they are only guesses based on what little we know about the universe. For example, they estimate there are 200 million trillion stars in the universe, many of which have planets. Some estimate there are 21.6 sextillion planets in the observable universe. And how many black holes, comets, asteroids, and other cosmic objects exist? No one knows.

Website: More information about these estimates can be found on the following website: https://skiesandscopes.com/how-many-planets

Despite the complexity and immenseness of the universe, Jesus knows the names of the countless stars.

> And as the stars of the sky cannot be counted. (Jeremiah 33:22 NLT)
>
> Lift up your eyes on high And see who has created these stars, The One who leads forth their host by number, He calls them all by name; Because of the greatness of his might and the strength of his power, Not one of them is missing. (Isaiah 40:26 NASB)

### Stars Differ by Type and Size

The Father designed and Jesus created stars of different types and sizes. We see this in the following verse that refers to their created glory:

> The sun has one kind of glory, while the moon and stars each have another kind. And even the stars differ from each other in their glory. (1 Corinthian 15:41 NLT)

Science agrees with the Bible about stars, as it does on many other topics. For example, they agree that stars vary in size, color, and brightness. And that there are red dwarf stars, white dwarf stars, brown dwarf stars, red giant stars, and neutron stars.

Website: You can read more about stars on the following website: https://science.nasa.gov/universe/stars/types/

Reminder: Even though science and the Bible agree on the existence of so many different types and sizes of stars, the above article purports the theory that everything in the universe evolved over billions of years.

### Composition of the Universe

Scientists and creationists believe the universe is made up of space, time, matter, and energy. They agree that these did not always exist. They also agree that there was a beginning to the universe. They agree the universe consists of stars, solar systems, galaxies, nebulae, black holes, and other cosmic phenomena. Many believe everything in the universe is moving. The prominent area of disagreement is about how it came into existence.

### Planet Earth Is in the Universe

Of course, Earth is part of the universe. This seems like a logical statement. We sometimes use the phrase "outer space" to refer to the universe that exists beyond Earth. But where does outer space begin? Scientists say that outer space starts 62 miles above the Earth's surface.

That's not very far, considering the universe is so big we cannot conceive of its actual size. No wonder a spacecraft takes a short trip from Earth to reach outer space. For comparison, Louisville, Kentucky, is 68.17 miles from Danville, Kentucky. Driving distance is 82.8 miles. You can drive that distance by car in an hour and a half.

Website: The following website tells us that outer space begins just 62 miles above the surface of the Earth: https://science.nasa.gov/exoplanets/what-is-the-universe/

## CREATIONIST'S VIEW OF THE UNIVERSE

Please remember that as creationists, we know that the Creator God made the universe in all its glory and majesty to reflect his glory and majesty. Even the fictional constellations reflect his glory.

> Who alone stretches out the heavens And tramples down the waves of the sea; Who makes the Bear, Orion and the Pleiades, And the chambers of the south; Who does great things, unfathomable, And wondrous works without number. (Job 9:8–9 NASB)

> On the glorious splendor of Your majesty And on Your wonderful works, I will meditate. (Psalm 145:5 NASB)

Our Creator God deserves all praise and glory for what he did in creating the universe and everything else.

> Let the heavens and the earth praise him, along with the seas and everything that swims in them! (Psalm 69:34 NET)

> The heavens are telling of the glory of God; And their expanse is declaring the work of His hands. (Psalm 19:1 NASB)

The universe and Earth reflect the glory and majesty of God, so no one can say he does not exist.

> For ever since the world was created, people have seen the earth and sky. Through everything God made, they can clearly see His invisible qualities—His eternal power and divine nature. So they have no excuse for not knowing God. (Romans 1:20 NLT)

The majestic and glorious nature of the universe testifies it came from the hand of the Creator God. The nature of the universe does not need to be explained as having been created. This is self-evident by its glory and majesty.

## CREATION OF EARTH'S SOLAR SYSTEM

When the Father God designed our solar system, he did so with precision and complexity. Jesus created everything according to his design and plan. Everything stays in its proper place. Jesus keeps it all synchronized so Earth doesn't crash into other planets and moons and self-destruct.

Website: More information about the planets and moons in our solar system can be found on the following website: https://theplanets.org/moons

Take a guess and write down how many planets and moons exist in our solar system. You may be amazed if you don't know already. The answer is at the end of this chapter.

## WAS THE UNIVERSE CREATED MATURE?

An article on the Christian creation website answersingenesis.org says that Jesus created the universe to be mature and fully functional. *In the beginning God created the heavens and the earth* (Genesis 1:1 ESV). These Young Earth Creationists believe the universe was created fully mature and complete with all the stars, solar systems, galaxies, nebulae, black holes, and all other objects that exist today. This idea is part of their theory about the age of the universe, Earth, and people being six thousand years old. We know that Jesus spoke the universe into existence out of nothing *(Ex Nihilo)* and continues to sustain it as the Father God desires. However, there is debate even among creationists about whether the universe was created fully mature or whether it was created and evolved to its current (mature) state.

Website: More information about their theory can be found on the following website: https://answersingenesis.org/astronomy/age-of-the-universe/mature-for-her-age

Instructional Comment: As a creationist, I believe the universe was instantly created fully mature and complete (Genesis 1:1). However, I can see the remote possibility that Jesus, instead, might have created and guided the evolution of the universe's maturity over millions or billions of years. Some Christians believe that he created the universe with a Big Bang. When I say this, I am not saying I (or they) believe in the theory of evolution, which I do not. The chapter, "Delving Deeper: The Big Bang Theory" reviews this theoretical idea of a Big Bang startup.

---

> Worldview: Some people believe the universe was created instantly to be mature and complete. Another theory is that was created by Jesus using the Big Bang Theory, who guided it over billions of years to its current state.

---

## THE UNIVERSE IS KEPT SECURE

Galaxies (like our Milky Way) contain billions of stars, their related solar systems, and gases and dust. Astronomers theorize they are moving through each other at incredible speeds with synchronized motion. How is it that they don't collide and destroy each other? The answer is that Jesus keeps the entire universe secure, just as the Father desires. Even when new stars are born, and old stars die, Jesus keeps everything in its proper place by his word.

> He existed before anything else, and He holds all creation together. (Colossians 1:17 NLT)

# SCIENTIFIC DISCOVERIES: JESUS SUSTAINS UNIVERSE

New theories from the sciences provide another perspective on how Jesus may have created the universe.

## The "God Particle"

The long-sought "Higgs boson" ("God particle") was confirmed to exist in 2012 at the Large Hadron Collider. They say the God particle helps us understand why things with mass exist, such as planets, galaxies, and even people. Mass is how much matter (substance) something has. It can be measured on Earth by how much something weighs. This God particle helps give mass to all elementary particles that have mass, such as electrons and protons of atoms. Photons that make up light do not have mass and are not affected by the God particle. Studies continue to develop our scientific understanding of these and how they affect the universe.

Website: More information about the discovery of the God particle can be found on the following website: https://www.nationalgeographic.com/science/article/120704-god-particle-higgs-boson-new-cern-science

Note: For creationists, this is another scientific discovery that supports the biblical account that Jesus Christ created and sustains everything in the universe and on Earth according to the Father's design and will.

## Dark Matter and Dark Energy

Cosmologists investigating the universe cannot explain why our universe, even planet Earth, remains intact. What makes them stay where they are? For example, why does the Earth remain in orbit around the sun? Why does our sun stay in orbit in our galaxy? To explain this unknown phenomenon, cosmologists theorize the existence of an invisible substance, "dark matter." Because it cannot be seen, they propose it exists by the observable evidence of an effect on planets, solar systems, galaxies, and other cosmic objects. That universal effect is thought to be gravity. Scientists speculate that dark matter is like a web throughout the universe that holds it together. They think dark matter accounts for about 27 percent of the universe.

Cosmologists theorize the universe is expanding faster and faster. But what makes this happen if the universe is only composed of matter (including dark matter)? There is nothing visible to explain this expansion. They believe this results from invisible energy called "dark energy." All energy is invisible, but we can see the effects of it. When you turn on a light in your home, you see the effect of energy. Cosmologists theorize that dark energy accounts for about 68 percent of the known universe. Therefore, dark matter and dark energy compose 95 percent of the universe. This means that all solid matter, such as planets, suns, asteroids, and so on, account for only 5 percent of the universe. That is an incredible thought!

Website: More information about dark matter and dark energy can be found on the following website: https://spaceplace.nasa.gov/dark-matter/en/

**We Are Still Learning about the Universe**

There is much about our universe that we still do not understand. Scientists develop theories based on observable evidence to increase our understanding and eliminate unknowns. Since the Bible is not a book on science, many scientists do not think it provides explanations related to science. But there are scientific facts in the Bible. And in the case of the universe, there is a biblical fact that explains why everything in the universe remains as it should. Jesus not only created the universe, but he also sustains it. He holds it all together according to the Father's design and will. Jesus is part of the Triune God and is supernatural. All that he does is beyond the natural world he created. How he is able to speak things into existence and sustain them in existence is beyond our human imagination. If dark matter and dark energy exist, it is because Jesus created them to do what they are doing.

Mystery: How is Jesus sustaining and holding the universe together?

## ASTOUNDING 2024 SPACE TELESCOPE PHOTOS OF THE UNIVERSE

The new NASA James Webb Space Telescope captured images of 19 spiral galaxies. It was sent into space two years previously with new, highly sensitive technologies, allowing it to see farther into space with more detail than the Hubble telescope. They say it provides more knowledge about the evolution and structure of galaxies, star formation, the life cycle of stars, black holes, and more. The scientists say a fixed longer-term focus on a tiny portion of the universe allows the telescope to continue to trace light from the universe's faintest and most distant stars and galaxies. They hope to discover the theoretical first galaxies formed by the Big Bang.

The website below offers breathtaking photos of a small portion of the universe taken by a new telescope. The scientists also express their ideas about the universe's evolution. But we know that our Triune God designed and created the universe and everything else, so it did not evolve. As you look at these photos, observe the majesty and glory of the Creator God and marvel at His handiwork!

Website: Please take the time to look at the incredible photos of the universe taken by this extraordinary new space telescope. You will be amazed at how wonderfully made the universe is. https://www.bbc.co.uk/news/resources/idt-611525eb-3a0c-4a68-bf54-485df138b6f6

## "HOW BIG IS OUR GOD" ANIMATION

A website with a video by Europa Technologies animates the largest-sized objects in the universe and the smallest-sized objects in the human eye. Of course, our Triune God created everything in the universe and the human body. I call the video "How Big Is Our God" since it shows how incredibly complex and intricately designed the universe and human eye are. I am not sure if the creators of this video are Christians, but the images support the belief in an Intelligent Designer of all things.

Website: You can watch this amazing animation on the following website: "Cosmic Eye 'Louise.' A Zoom Journey from the Universe to the Sub-atomic." https://www.europa.uk.com/cosmic-eye-louise.

### Answer: How Many Planets and Moons in Our Solar System?

There are eight planets and five dwarf planets. In order of their closeness to our sun, they are Mercury, Venus, Earth, Mars, Ceres, Jupiter, Saturn, Uranus, Neptune, Pluto, Haumea, Makemake, and Eris being the furthest from the Sun. Pluto was reclassified as a dwarf planet due to its small size and other factors. You may be surprised to learn that our thirteen planets and dwarf planets have 181 known moons. Jupiter has the most at 67, while Saturn is second at 62. Four planets do not have moons: Mercury, Venus, and the dwarf planets Ceres and Makemake. Of the eight planets, Mercury, Venus, Earth, and Mars have a solid surface with a dense, rocky composition. Jupiter, Saturn, Uranus, and Neptune are massive and composed of gaseous substances.

Instructional Comment: Plutonium and Neptunium are synthetic elements in the Periodic Table. They were named after the planets Pluto and Neptune. Uranium is a natural element in the Table and is named after the planet Uranus. Mercury is a natural liquid metal in the Table. It is believed to be named after the mythical god Mercury, with an ancient link to the planet Mercury.

## CHAPTER QUESTIONS

DQ1: Cosmologists estimate there are 200 million trillion stars in the universe. Some estimate there are 21.6 sextillion planets in the observable universe. Despite the unknown immenseness of the universe, Jesus knows the names of the countless stars (Jeremiah 33:22). What do you think is the purpose of Jesus knowing the names of these myriad of stars?

DQ2: In what ways are the heavens "telling of the glory of God; And their expanse is declaring the work of His hands?" (Psalm 19:1 NASB)

DQ3: Creationists believe the universe was instantly created fully mature and complete (Genesis 1:1). Why do you think it might be possible that Jesus created and guided the evolution of the universe's maturity over millions or billions of years, perhaps starting it with a singularity and expanding it with the idea of a Big Bang?

DQ4: Describe the theories about dark matter and energy and why they comprise 95 percent of the universe.

## DIG DEEP

DD1: Scientists and creationists believe the universe is made up of space, time, matter, and energy. They agree that these are the basis of stars, solar systems, galaxies, nebulae, black holes, and other cosmic phenomena. Explain how space, time, matter, and/or energy affect our solar system.

## PERSONAL APPLICATION

PA1: Why not take time now to express appreciation to the Creator God for his visible majesty and power in the universe?

PA2: How do these theories about the God particle, dark matter, and dark energy help you understand how inventive and almighty the Creator God is?

# PART J

## Delving Deeper into the Universe

# 34.

# Delving Deeper: Did God Create Aliens?

Important: Delving Deeper chapters contain material that may be difficult to understand. I suggest you spend extra time studying it and ask for the Holy Spirit's help. The Father and Jesus sent him to live within Christians as their Teacher, Guide, and Helper.

This book is about creation and evolution. Chapters provide biblical and scientific research to help increase our understanding of these topics. This chapter on aliens (extra-terrestrial life) provides a broad view of the theory that intelligent, sentient life might exist on other planets and galaxies. Much of this chapter is hypothetical since little evidence supports the theory of alien life. Of course, if aliens exist, they do so only because the God of the Bible designed and created them. (They did not evolve independently of the Creator God.)

---

> There is little evidence to support the theory of alien life on other planets. Of course, if aliens exist, they do so only because the God of the Bible designed and created them. (They did not evolve independently of the Creator God.)

---

## GOLDILOCKS ZONES IN THE UNIVERSE

NASA's Transiting Exoplanet Survey Satellite recently spotted a planet in a "habitable zone" that might have the potential to support life. A habitable zone is an area of space around a star where liquid water can exist on a planet without freezing or boiling. The term "Goldilocks Zone" means it's an area of space that is not too hot or too cold to potentially sustain life.

The recent planet discovered in this zone is designated TOI-715b. Its star is a red dwarf that is smaller, cooler, and emits much less light than our sun. However, the planet is large, about 1.5 times the diameter of Earth. Scientists seek a smaller Earth-sized planet that might accompany it and support life with water. In this article, there is no mention of creation or that God may have created habitable planets.

Website: The following NASA website has more information about this potentially habitable planet: https://www.space.com/exoplanet-super-earth-habitable-zone-tess

Note: The term Goldilocks Zone is derived from the fable "Goldilocks and the Three Bears."

## WHY WOULD GOD CREATE ALIENS?

When considering the possibility of extraterrestrial life, the first questions I think of are, "Why would God create aliens on other planets?" "What would be his reason and purpose for their existence?" To answer this, let's look through the lenses of science and the Bible.

## SCIENCE AND ALIENS

Studying science and the potential of alien life without considering the God who created everything results in partial answers. If aliens exist, the God of the Bible would have created them.

Following are some questions from a scientific perspective about the possibility of the creation of extraterrestrial life:

- To create aliens, did God use the same genetic material and genetic templates he used on Earth?
- Since science theorizes that the 92 natural elements of our Earth's Periodic Table exist in space, do they exist on habitable planets?
- If he created life on other planets, were they human (like us) and made from natural elements?
- Did he create oxygen-based plant and animal life to co-exist with them?
- The Earth was formless and void of life after the creation of the universe. Was this true of other planets on which God might have created life?
- If aliens exist, do they exist in a different universe or part of our universe?

Note: The above thoughts are purely speculative.

Definition: *Aliens* are theorized to be intelligent, sentient beings who exist on other planets and galaxies.
*Extraterrestrial life* is another name used for aliens. The term means life that might exist outside the Earth and its atmosphere.
*Alien humans* are theoretically aliens designed and made like Adam and Eve from the 92 natural elements on Earth who need an oxygen-rich environment to exist.
*Non-human aliens* are theoretically aliens designed by the Father God who need gasses other than oxygen-rich environments to exist and might not be made of the 92 natural elements of Earth.

## EARTH'S ATMOSPHERE NEEDED FOR HUMAN LIFE

When considering how God might have created life on other planets, we should consider how the atmosphere around the Earth enables life to exist.

The following are some additional scientific facts about breathable air in the Earth's atmosphere. Our atmosphere is composed primarily of gasses, with minute amounts of particulate matter.

- Dry air of the Earth's atmosphere (by quantity of molecules) is composed of 78.08 percent nitrogen, 20.95 percent oxygen, 0.93 percent argon, 0.04 percent carbon dioxide, and small amounts of other trace gases.
- The amount of water vapor in the air varies depending on its location: 1 percent water vapor at sea level and .04 percent throughout the entire atmosphere, and 4 percent in tropical climate locations.
- Altitude affects the gaseous composition of air, its typical temperature, and atmospheric pressure.
- Gravity creates atmospheric pressure and prevents Earth's gasses from being vented into space.
- Air suitable for plant photosynthesis and animal respiration is found up to 7.5 miles from the Earth's surface.

Note: Theoretically, oxygen-based life on other planets would require a similar atmosphere to Earth's.

### Oxygen: Basis for All Life on Earth

The basis of all plant and animal life on Earth is the essential gas, Oxygen (O). We already know that the human body is comprised of 60 percent water ($H_2O$) and 65 percent oxygen. So, if aliens are humans like us, they must have an oxygen and water-rich natural environment on their planet. If not, the facts about Earth's atmosphere do not apply to them. Therefore, some other gasses are needed to support non-human alien life. Gasses toxic to oxygen-based life on Earth are chlorine, fluorine, carbon monoxide, carbon dioxide, and many others in medium to high amounts. Is it possible that Jesus created non-human alien life on other planets that are rich in one or more of these Earthly toxic gasses?

## NON-OXYGEN BREATHING ALIENS?

Could aliens exist that thrive in environments that are not water and oxygen-rich, such as chlorine or fluorine? God's imagination and creativity are far beyond the capability of our human minds to fathom. He can do anything he chooses. Science fiction movies and some accounts of seeing aliens on Earth depict non-human aliens as small, green, and frail with big heads that cannot live in our oxygen-rich environment. Maybe their skin color is depicted as green because of the non-oxygen gasses within them? This is just wild speculation with no basis in truth.

Instructional Comment: If intelligent, sentient beings exist on other planets of any gaseous substance and physical form, the God of the Bible designed and created them according to his purposes.

## WHAT ABOUT UNIDENTIFIED FLYING OBJECTS?

You have likely read stories and even seen video footage of purported unidentified flying objects sighted in our global skies. Some military and commercial pilots claim to have seen them following or hovering near planes. Some credible scientists claim to have seen the wreckage of alien spacecraft and even the dead bodies of aliens. So, what is the truth? Have aliens from other planets or galaxies visited Earth? Or are these purported extraterrestrial experiences simply misinterpreted natural phenomena?

Definition: *Unidentified Flying Object* is a term used to label any object seen in our global skies that could not be clearly identified.
*Unidentified Anomalous Phenomena* has replaced the Unidentified Flying Object term in an attempt to remove the conspiracy stigma that might prevent reportings of purported sightings to the government. They want to know what people believe they have witnessed so they can investigate them.
*Identified Flying Objects* are Unidentified Flying Objects that have been identified.

Following is a true story about witnessing a flying saucer:

> In the mid-1960s, a family member and several friends were returning to their hometown late at night after a movie in a nearby town. As they traveled the road home, they heard a whirring noise above them. Looking up, they saw an unmistakable small flying saucer moving slowly about 20 or 30 feet above the ground. This scared them so much that they vowed never to tell anyone. They were afraid they would be labeled as crazy conspiritists. Over 20 years later, an article in the other town's newspaper showed a picture of this same small flying saucer with the engineer who built it. It turned out he was a local university professor who built it as an experiment. It seems its velocity and height of ascent were limited based on known engineering and aeronautical technologies. So, yes, flying saucers can be built and exist on Earth, but only by humans and with limited aeronautical capabilities.

## NAZCA LINES, CAVE DRAWINGS, AND DEMONS

Is there really any evidence that extraterrestrial life exists? Ancient petroglyphs and cave paintings seem to depict aliens and their spacecraft. The Naza Lines seem to indicate the need for aliens to have created the lines.

## Nazca Lines

There are over 700 drawings ("geoglyphs") on the Peruvian landscape referred to as the "Nazca Lines of Peru." Some are relatively small, while others are large and extend over significant distances. They are estimated to have been made from 500 BC to 500 AD. Interestingly, some are only visible in their entirety from a high elevation. Why create these incredibly detailed images, some of which can only be seen from great heights? No aircraft existed at that time. So, did aliens create these as a way to declare their existence to people on planet Earth? Or, as some archaeologists theorize, did the Nazca people create them for religious purposes?

## Cave Drawings

Cave drawings around the Earth depict saucer-like crafts and aliens on Earth. These artistic renditions of aliens, however, don't look like us. Where did these primitive people get the idea and images of aliens visiting Earth? Have people been imagining aliens visiting Earth for thousands of years, or have they been here?

Website: More information about these cave drawings can be read on the following website: https://www.ancient-code.com/5-ancient-petroglyphs-cave-paintings-that-depict-ancient-aliens

## Demon Manifestations?

People tend to see what they want to see. If they want to see unidentified flying objects, they will see them regardless of what is real. Some people wonder if some experiences of alien spacecraft and alien beings have come from demons. Demons have been around since before creation. They know people, their fallen sinful nature, and their tendencies. They can put thoughts into people's minds to lead them astray from God (Mark 8:29–33). (However, demons cannot read people's minds.)

Alien abductions have been reported by people around the world. Most are experienced in dreams. Some of these dreams are pleasant, while others are horrifying. We know the devil and his demons are liars and deceivers. They want to confuse people and keep them occupied with anything that captures their attention. I wonder if the mass fascination with the possible existence of aliens has accomplished this diversion from likely reality to some extent.

# CREATION, CHRISTIANITY, AND ALIEN LIFE

*The Created Cosmos* was written by Danny R. Faulkner in 2016 and published by Master Books. Mr. Faulkner has graduate degrees in physics and astronomy and is a member of the Creation Research Society. He is a conservative, evangelical Christian and a scientist. His book focuses on what the Bible teaches about astronomy and the universe. He reviews misconceptions about what people think the Bible teaches about astronomy, the universe, and potential life on other planets. He says there is no biblical or scientific proof that extraterrestrial beings exist. He also says most unidentified flying objects have rational explanations. Even though he believes science supports the theory of Earth-like planets in Goldilocks Zones, he says no planet has been found with human-like life. *The Created Cosmos* is a good book for anyone who wants to study astronomy and the universe from a biblical and scientific perspective.

## BIBLE AND ALIENS

The most challenging question about the possibility of extraterrestrial life is, "If the Triune God created aliens to be like humans on other planets, were they made in his image?" This thought brings up some speculative theological considerations, such as the following:

- Were theoretical aliens humans made to be eternal and perfect like Adam and Eve?
- Were they made, male and female, in the image of the God of the Bible?
- Did God create a Garden of wonderful, perfect beauty for them to enjoy and work in?
- Did he create seed-bearing plants and trees for food?
- Did he plant a Tree of Life and a Tree of the Knowledge of Good and Evil?
- If so, did God allow them to be tempted by the devil to see if they would obey him?
- Did they believe the devil's lies and deception and lose their eternal, perfect nature?
- Did their sin of rebellion cause all their planet's creation to be corrupted?
- Did Jesus Christ come to their planet as their Savior and Lord (to be killed and rise from the dead) to save aliens from their sins against God?
- Did they have someone like Moses and others to write books for a Bible like ours based on their knowledge of God and life on their planet?
- Do the devil and his demons roam their planet looking for someone to devour?
- Will God create a new universe and planet for them at the end of time, or will they be part of ours?

Instructional Comment: Our Almighty Triune God could design and orchestrate whatever he wants for other planets in his universe. The Bible does not tell us whether people were created to exist on other planets. If you are interested in this possibility, you can only make assumptions and speculate about what might be true. I suspect the biblical account of the creation is unique to planet Earth and that aliens in any form do not exist on other planets.

## CHAPTER QUESTIONS

DQ1: Why do you think God might have created human-like beings (aliens) on other planets?

DQ2: If there is life on other planets, why is it possible Jesus used the same genetic templates of creation kinds to create life on them?

DQ3: Why does the theory about the possible existence of "just right" planets to sustain life seem plausible?

DQ4: Why do you think some purported extraterrestrial experiences might be misinterpreted natural phenomena rather than extraterrestrial?

## DIG DEEP

DD1: Summarize why the questions from a science perspective about the possibility of alien life might lead you to believe human-like life could exist on other planets.

## PERSONAL APPLICATION

PA1: Why does the theory that alien abductions not frighten you as a follower of Jesus Christ?

PA2: What would you want to say to a true alien if you met one?

# 35.

# Delving Deeper: The Big Bang Theory

Important: Delving Deeper chapters contain material that may be difficult to understand. I suggest you spend extra time studying it and ask for the Holy Spirit's help. The Father and Jesus sent him to live within Christians as their Teacher, Guide, and Helper.

Some people and scientists believe the universe evolved about 13.7 billion years ago from a theory called "The Big Bang."

Website: The website *space.com* provides an interesting theory of how scientists think the Big Bang occurred 13.7 billion years ago: https://www.space.com/25126-big-bang-theory.html

Scientists supporting the Big Bang Theory propose the universe's ancient age by working backward from what they believe is true about the current universe. They theorize the universe started with a singularity (single point of matter and energy) that was infinitely hot and dense. This single point inflated and expanded at incredible speeds over a period of 13.7 billion years to become the universe today. These scientists believe the universe is still expanding. Mathematical formulas, observations of the universe, and theoretical models form the basis for the Big Bang Theory. Many modern scientists do not accept the Big Bang Theory as proof of the origin and current state of the universe.

---

Mathematical formulas, observations of the universe, and theoretical models form the basis of the Big Bang Theory. Many modern scientists do not accept the Big Bang Theory as proof of the origin and current state of the universe.

---

Note: Most evolutionists who support the Big Bang Theory do not believe in an Intelligent Designer and Creator God.

## EARLY SCIENTIFIC SPECULATION

The following are brief descriptions of early twentieth-century theories about the nature of the universe. I am omitting the names of most of the scientists since many readers may not be familiar with them.

Definition: *Cosmology* is the study of the universe and how it functions.
*Cosmos* is another name for the universe.

### Universe Expanding Theory

The following are three of the earliest theories about the universe expanding:

> 1924: After observing the universe through telescopes, some astronomers began to theorize that the entire universe was expanding at an incredible rate. This means all the stars, their solar systems, their galaxies, and all other objects in the universe are moving and not static as previously thought. They believed the universe could be measured and concluded these celestial bodies were all moving farther away from each other. The question that arose for some was, "Will the universe continue to expand outwardly forever or stop at some point (whatever that point might be)?"
>
> 1929: Edwin Hubble ("Hubble Telescope" fame), using the Hooker Telescope, published similar findings about the expanding universe.
>
> Redshift and Blueshift (lightwave frequencies)
> Edwin Hubble theorized redshift and blueshift as a way to determine if a cosmic object was moving closer to or farther away from Earth. Astronomers use these lightwave frequencies as evidence to support the theory that the universe is expanding. They believe the universe's expansion stretches the wavelengths of light traveling through it. Redshift represents a longer (or stretched) light wavelength as objects move away from Earth. The blueshift represents a shorter light wavelength as objects move closer to Earth. Redshift and blueshift correspond to distances from Earth in light years. Some scientists today use this theory to provide evidence that the Big Bang Theory correctly describes the origin of the universe.

Website: More information on redshift and blueshift is available on the following science website: https://earthsky.org/astronomy-essentials/what-is-a-redshift/

### Big Bang Theory

Following is one of the earliest theories about the Big Bang, along with two more modern theories about its existence:

> 1931: Father Georges Henri Joseph Édouard Lemaître, a Belgian Catholic priest, postulated the entire universe was expanding from a single point of origin. He believed that running the outward expanding universe of space and time backward would lead to that initial single point. He theorized it was a point of infinite density and temperature. His theory became known as "The Big Bang Theory." Later, this

single point from which the entire universe was theorized to expand was called a "singularity."

1964: "Background radiation" was thought to have been discovered in the universe in the 1960s. Scientists explain its existence as the result of the Big Bang Theory. They believe the theoretical early universe after the Big Bang was intensely hot and cooled off as it expanded. Therefore, the universe would be filled with leftover heat from the cooling process. This leftover heat was thought to be background radiation. This is called the "cosmic microwave background." Another way of stating this theory is that the cosmic microwave background is theorized to be the remnant of the first light wave that traveled freely (without obstruction) throughout the universe as it expanded.

1990s: Scientists said NASA's Cosmic Background Explorer satellite had also measured the microwave radiation from the early universe. Their observations and theoretical conclusions are said to support the Big Bang Theory.

Website: More about the evolution of the universe and the Cosmic Background Explorer satellite can be found on the following NASA website: https://www.nasa.gov/specials/60counting/universe.html

## SOME DOUBTING SCIENTISTS

Some scientists agree that the universe is moving but doubt it is expanding. They believe that galaxies are moving away from each other. Their theory purports that when something grows, it must expand into a larger space than it currently exists within. They equate expanding with growing larger. (For a comparative example, use a foam spray and see how much greater its volume (space) expands.) These scientists theorize there is no additional space for expansion within the universe's present space. Therefore, the universe cannot expand into space that isn't there. As a result, the universe must be finite, not infinite.

Mystery: Is the universe infinite or finite? Is it moving or static?

Instructional Comment: The Big Bang Theory is highly controversial, with little agreement about its factualness, even among scientists who are evolutionists.

## THE UNIVERSE IS FINITE, NOT INFINITE

As you have read, many scientific theories purport the universe is expanding from a beginning point (singularity). If true, the universe is not infinite since it has a beginning. There is also speculation among scientists about whether the universe has a physical termination point.

Infinity has no beginning and end. Only the God of the Bible is infinite and has no beginning or end. An infinite God could create an infinite universe, but because he created it, it had a beginning. We also see in the Bible that the current universe will cease to exist when God destroys it with fire (2 Peter 3:10–13; Revelation 20:11). This occurs just before Judgment Day. In Revelation 21:1, we read that God creates a new heavens (universe) and Earth for the followers of Jesus to live with God the Father, Jesus, and the Holy Spirit eternally. Therefore, the universe is finite, not infinite.

## SIZE OF THE UNIVERSE

It's impossible to know how big the universe is. We can only make assumptions and theorize. Some creationists believe the universe was designed and created full and complete, regardless of its size. They believe it's not expanding but may be moving. Others think the universe may have been designed and created by our Creator God using a singularity and the Big Bang Theory. (Only he knows how the universe was designed and created.)

### Evolutionist's View of the Universe's Size

Recent observations of small portions of the universe through powerful telescopes enable estimations (guesses) based on observable evidence and mathematical calculations about how large the *visible universe* may be. Scientists speculate that it extends 47 billion light years in every direction from the focal point from which they view it. The issue about its size hinges on whether it has a physical end (not a biblical end). If the universe has a physical end, where is that end, and in what way does it end? Is there an edge or boundary that could be reached? Or does it continuously loop back onto itself when it reaches its circumference and then continue moving within itself? These are theories since it's impossible to know how big the universe is and ho wit functions.

Website: More about these theories can be read on the following website: https://nautil.us/could-the-universe-be-finite-466593

### Young Earth Creationist's View of the Universe

Answers in Genesis is a Christian website with many articles and videos supporting the biblical creation account. Ken Ham and others provide evidence that the universe was created and did not evolve. One article identifies fourteen natural phenomena that conflict with the supposition that the universe evolved over billions of years.

Website: More information about the age of the universe in Answers in Genesis can be found on their website: https://answersingenesis.org/astronomy/age-of-the-universe/
Note: As a Young Earth Creationist, Ken believes the universe and Earth are about six thousand years old. He believes that Jesus instantly created the universe to be completely mature and fully functional so that it did not evolve or grow. This means he doesn't believe in the Big Bang Theory as a possible means by which the Triune God may have kicked off the universe's creation 13.7 billion years ago. The size of the universe is not relevant since what exists was created to be fully mature and functional.

## CHAPTER QUESTIONS

DQ1: Why do you think many scientists today do not believe the Big Bang Theory adequately explains how the universe came into existence?

DQ2: Does the doubter's logic about an expanding universe seem sound to you? Do you agree or disagree with it?

DQ3: Explain why the universe might not be infinite.

DQ4: How do the evolutionist and Young Earth Creationist views of the universe's size differ?

## DIG DEEP

DD1: Explain how the theory of redshift and blueshift lightwave frequencies explain the movement of a cosmic object in relation to Earth

## PERSONAL APPLICATION

PA1: Explain why the nature of the universe's size, finiteness, and motion doesn't affect your belief in the Creator God of the Bible.

# PART K

Creation of New Heavens (Universe) and Earth

# 36.

# Creation of New Heavens (Universe)

## ORIGINAL HEAVENS AND EARTH WERE PERFECT

The original heavens and Earth were created perfect, without defects or the existence of sin (Genesis chapters 1 and 2). The Triune God is perfect, so everything he did in creation would have been perfect. Therefore, his initial creation had no death, pain, suffering, toil, misery, or sin.

> But you are to be perfect, even as your Father in heaven is perfect. (Matthew 5:48 NLT)

> For I will proclaim the name of the LORD; ascribe greatness to our God! The Rock, his work is perfect, for all his ways are justice. A God of faithfulness and without iniquity, just and upright is he. (Deuteronomy 32:3–4 ESV)

How wonderful, beautiful, and awe-inspiring were the original universe and Earth before being corrupted by the rebellion of Adam and Eve against their Creator God!

## ORIGINAL HEAVENS AND EARTH BECAME CORRUPTED

When Adam and Eve defied God and rebelled against his Lordship, loving-kindness, and provisions, the perfect Adam and Eve became corrupted in their human nature. As a result, all of God's original creation became corrupted. The universe and the Earth are now corrupted in their very nature. Death, misery, toil, pain, and suffering are a natural part of all life on Earth.

> For the anxious longing of the creation waits eagerly for the revealing of the sons of God. For the creation was subjected to futility, not willingly, but because of Him who subjected it, in hope that the creation itself also will be set free from its slavery to corruption into the freedom of the glory of the children of God. (Romans 8:19–22 NASB)

## MILLENNIAL REIGN: RENEWED HEAVENS AND EARTH?

By the end of the seven-year Tribulation, the Earth will be completely destroyed by both God's wrath and the wars and destruction of people. We see this in the book of Revelation, chapters 6 through 18. The Earth will not be able to sustain human or any other life. There may be no oxygen, no freshwater, no ozone layer to protect from the sun's ultraviolet radiation, and no food from the oceans, rivers, or crops from the land. Jesus will need to renovate this destroyed Earth for His Millennial Reign.

Note: I don't see Jesus destroying or obliterating this uninhabitable Earth. He does that later after his Reign.

### Who Will Be on the Renewed Earth?

In Revelation, chapters 19 and 20, we read about the Millennial Reign of Christ on Earth. We see that all the saved who were martyred by the antichrist and his armies will reign on Earth with Jesus for a thousand years. (I assume they will be given their eternal resurrected bodies.) Some think all believers from the beginning of time to the end of the Tribulation will reign with Jesus, not just the Tribulation saints. Others believe there will be some new believers who will be alive at the end of the Tribulation. If so, will they be given resurrected bodies or go into the Reign with their human, mortal bodies?

It appears believers will have children (Revelation 20:7–10) who may or may not be given resurrected eternal bodies during the Reign. Who is reproducing these children? There are many questions about this Millennial period for which we do not find answers in the Bible.

> Mystery: To what extent will the Earth need to be renovated to be habitable for people and other life at the start of the Millennial Reign? Will some people alive from the Tribulation be there in human form without resurrected bodies? Will animals be there? Will people and animals need food and water?

Instructional Comment: I believe the massive destruction of Earth will require it to be renewed to support human and other life. I find no verses in the Bible to support this renewal supposition. It seems common sense to me. This is just my opinion.

## HEAVENS AND EARTH DESTROYED BY FIRE

God tells us very clearly that he will destroy this current corrupted universe and Earth by fire.

> But by His word the present heavens and earth are being reserved for fire, kept for the day of judgment and destruction of ungodly men. (2 Peter 3:7 NASB)

> But the day of the Lord will come like a thief, in which the heavens will pass away with a roar and the elements will be destroyed with intense heat, and the earth and its works will be burned up . . . . the heavens will be destroyed by burning, and the elements will melt with intense heat! (2 Peter 3:10–12 NASB)

# CREATION OF NEW HEAVENS (UNIVERSE)

*Destroyed*, Greek is *luō*; to loose, to release, to dissolve.

*Burned up*, Greek is *katakaiō*; consume wholly, burnt up utterly.

*Melt*, Greek is *tēkō*; to liquefy.

*Elements*, Greek is στοιχεῖον; One meaning is the natural, material elements of the universe (This would mean the 92 elements of the Periodic Table)

From the following verse, it appears the universe will dissolve and melt away into nothing from the intense heat that destroys it.

> And all the host of heaven *(stars, solar systems, and galaxies)* will wear away, And the sky will be rolled up like a scroll; All their hosts will also wither away As a leaf withers from the vine, Or as one withers from the fig tree. (Isaiah 34:4 NASB, author's emphasis)

*Wear away*, Hebrew is *mâqaq*; to melt, vanish.

*Withers*, Hebrew is *nabel*; to sink or drop down, fade.

## Which of the Biblical "Three Heavens" Are Destroyed?

It's difficult to find these in the Bible, but the three heavens are as follows:

- Atmosphere around the Earth where birds fly (Psalm 19:1)–Destroyed by fire
- Universe, or "outer space" (Psalm 8:3)–Destroyed by fire
- Heaven/Paradise, where the Father God and Jesus dwell (2 Corinthians 12:2)–Not destroyed; comes down to new Earth

## OUR SOLAR SYSTEM DESTROYED

Fire will destroy our sun, moon, and all the planets and their moons in our solar system. When God creates the new heavens (universe), he does not create a solar system for the Earth to spin around.

> Then I saw a new heaven and a new earth, for the first heaven and earth had ceased to exist, and the sea existed no more. (Revelation 21:1 NET)

> The city does not need the sun or the moon to shine on it, because the glory of God lights it up, and its lamp is the Lamb. (Revelation 21:23 NET)

Without the sun and moon for the new Earth, the following would be true:

- No time on Earth—no days, nights, seasons, or years
- No other planets to rotate around the sun
- No need for the Earth to be tilted 23.50 on its axis

Instructional Comment: The gravitational pull from the sun keeps the Earth in orbit around the sun. Without gravity, the Earth would spin off into the universe until it collided

## OUR MILKY WAY GALAXY DESTROYED

Our solar system is part of the Milky Way galaxy. So, it will likely no longer exist when God creates the new universe. That means its estimated over 200 billion stars will be burned up by fire.

Website: More information about our Milky Way Galaxy can be found on the following website: https://www.space.com/19915-milky-way-galaxy.html

Instructional Comment: Our Milky Way is a typical barred spiral galaxy. Some scientists estimate it has between 100 and 400 billion stars. Galaxies are physical structures in space. On the other hand, constellations are imaginary images based on the pattern of stars as seen from Earth.

Definition: A *Galaxy* contains stars, planets, interstellar gas, dust, and other cosmic objects held together as a structure by gravity (and dark matter?). Galaxies are classified by their shape as spiral, elliptical, and irregular.

> Mystery: How much of the immense universe will be destroyed by fire and replaced with a new universe?

### When Does God Destroy the Universe and Earth?

This fiery destruction occurs *after* the Tribulation and *before* the Great White Throne Judgment of all non-believers.

> Immediately after the tribulation of those days the sun will be darkened, and the moon will not give its light, and the stars will fall from heaven, and the powers of the heavens will be shaken. (Matthew 24:29 ESV)
>
> *Fall*, Greek is *piptō*; descend from a higher place to a lower, to fall, to be thrust down.
>
> *Powers*, Greek is *dunamis*; inherent power, ability to do something, power residing in a thing by virtue of its nature.
>
> *Shaken*, Greek is *saleuō* motion, to agitate or shake, to shake thoroughly.
>
> Then I saw a great white throne and Him who sat upon it, from whose presence earth and heaven fled away, and no place was found for them. (Revelation 20:11 NASB)
>
> *Place*, Greek is *topos*; a spot (generally in space), location (as a position).

## NEW HEAVENS AND NEW EARTH PROPHECY

Two thousand years after the first coming of Christ, the Holy Spirit gave Peter and John, Apostles of Jesus, a prophecy about the future new heavens (universe) and new Earth.

> But according to His promise we are looking for new heavens and a new earth, in which righteousness dwells. (2 Peter 3:13 NASB)

> Then I *(John)* saw a new heaven and a new earth; for the first heaven and the first earth passed away, and there is no longer any sea. (Revelation 21:1 NASB, author's emphasis)

God always does what he says he will do. This prophecy of new heavens and new Earth will occur in his timing.

## WHAT COMPRISES THE NEW UNIVERSE?

It seems clear from Scripture that the new universe will be located in what we think of as outer space (second heaven). How many galaxies, solar systems, stars, and planets will be created is unknown to us.

---

> It seems from Scripture that the new universe will be located in what we think of as outer space (second heaven). How many galaxies, solar systems, stars, and planets will be created is unknown to us.

---

Note: Heaven will not be destroyed since it's God's home. Instead, the Father and Jesus bring it down to the new Earth to continue to be the place for their thrones.

### What about Atmosphere and Gravity for the New Earth?

Believers will have eternal resurrected bodies on the new Earth. Therefore, there will be no need for oxygen in the atmosphere of the new Earth since people, animal life, and plant life will not need it to sustain their eternal lives. Their eternal bodies will not need anything to survive on the eternal new Earth. Water vapors (clouds) and water on the new Earth will also not be required.

Reminder: The basis of all plant and animal life on Earth is the essential gas, Oxygen (O). We also know that the human body is comprised of 60 percent water ($H_2O$) and 65 percent oxygen.

There will be no need for gravity to retain gasses on the new Earth in its atmosphere since there will likely be no gasses on or around it. Gravity (dark matter?) will also not be needed to prevent the new Earth from floating away from its new location in outer space.

Instructional Comment: I don't see a biblical or scientific reason for an atmosphere around the new Earth.

### How Many Stars and Planets in the New Universe?

No one knows for certain how many cosmic objects exist in the current universe. Scientists make their best guess, knowing they are only guesses based on what little we know about the universe. For example, they estimate there are 200 million trillion stars in the universe, many of which have planets. Some estimate there are 21.6 sextillion planets in the observable universe.

We can only guess what cosmic objects will be created for the new universe. Perhaps there will be new eternal galaxies with a multitude of eternal new stars and planets. Since there will be no more sin, death and destruction will no longer exist in the universe. That means the new eternal stars will not die. Some "black holes" in space are the result of a dying star. This will not occur. Dark matter and dark energy are theories. We don't know if they will created in the new universe.

## KNOWING THIS, HOW SHOULD CHRISTIANS LIVE?

How should Christians live if they really believed in the following prophecy of this fiery destruction?

> Since all these things are thus to be dissolved, what sort of people ought you to be in lives of holiness and godliness, waiting for and hastening the coming of the day of God, because of which the heavens will be set on fire and dissolved, and the heavenly bodies will melt as they burn! But according to his promise we are waiting for new heavens and a new earth in which righteousness dwells. (2 Peter 3:11–13 ESV)

> Just as He chose us in Him before the foundation of the world, that we would be holy and blameless before Him. (Ephesians 1:4 NASB)

Instructional Comment: To be holy means to be set apart for God and his use. God sees his children through the lens of Jesus's perfection as if we never sinned. Our response is to love and worship God with all our hearts, minds, and strength.

Important: Followers of Jesus Christ should view these verses as a promise, not a threat to cause them to fear. However, it is indeed a warning for everyone who refuses to accept the free gift of eternal life with God through Jesus Christ.

## CHAPTER QUESTIONS

DQ1: Why were the original heavens (universe) and the Earth perfect?

DQ2: Why do you think God will use fire to destroy the current corrupted Earth and universe?

DQ3: Why will people on the new Earth not need a solar system?

DQ4: Why do you think there might not be an (eternal) atmosphere around the new Earth?

## DIG DEEP

DD1: Why will the Earth not need to be tilted and rotate on its axis in the new heavens?

## PERSONAL APPLICATION

PA1: How should you live your life as a follower of Jesus Christ in light of this coming destruction and the creation of the new universe and Earth?

# 37.

# Creation of a New Earth

## GOD'S PURPOSE

What is God's purpose for the new Earth? I think it can be summarized in a simple way: *the new Earth will be a beautiful place for God to eternally dwell with his people.* It will be perfect because there will never again be sin, pain, suffering, or evil.

We see this in the following verses:

> And I saw the holy city, the new Jerusalem, coming down from God out of heaven like a bride beautifully dressed for her husband. I heard a loud shout from the throne, saying, "Look, God's home is now among His people! He will live with them, and they will be His people. God Himself will be with them. He will wipe every tear from their eyes, and there will be no more death or sorrow or crying or pain. All these things are gone forever." And the One sitting on the throne said, "Look, I am making everything new!" And then He said to me, "Write this down, for what I tell you is trustworthy and true." And He also said, "It is finished! I am the Alpha and the Omega—the Beginning and the End. To all who are thirsty I will give freely from the springs of the water of life. All who are victorious will inherit all these blessings, and I will be their God, and they will be My children." (Revelation 21:2–7 NLT)

---

Worldview: God's purpose for the new Earth can be summarized in a simple way: the new Earth will be a beautiful place for God to eternally dwell with his people.

---

### What Comes Down: New Jerusalem or All of Heaven?

Notice that Heaven in its entirety may not come down to the new Earth, only the "holy city, the new Jerusalem, coming down from God out of heaven."

> Mystery: What parts of Heaven don't come down, and what happens to them? What value or use are these without the Father and Jesus being there?

## RENOVATED OR UNIQUE NEW EARTH?

Will God simply take the current heavens and Earth and renovate them, remodel them as you would an old house? Since it will be perfect and eternal, he will create an entirely new Earth.

The Greek word for new in the following verses tells us that everything will be uniquely new, not renovated:

> But according to his promise we are waiting for new heavens and a new earth in which righteousness dwells. (2 Peter 3:13 ESV)

> Then I saw a new heaven and a new earth; for the first heaven and the first earth passed away, and there is no longer any sea. (Revelation 21:1 NASB)

> *New*, Greek is *kainos*; new (especially in freshness); did not exist before.

> *Passed away*, Greek is *aperchomai*; to go off (depart), pass away, be past.

### Created Out of Nothing (Ex Nihilo)

The uniquely new Earth (and universe) will be made from nothing (*Ex Nihilo*), just as the original universe and Earth were in Genesis 1:1. Perhaps Jesus will create a uniquely new Periodic Table with eternal elements. If so, they will not age, rust, or wear out as the corrupted natural elements do today. Have you seen rusted iron? If iron is created on the new earth, it will be eternal and never rust.

## CURRENT EARTH AND NEW EARTH COMPARISON

Everything about the new Earth will be different from the current Earth. For one thing, the Father God and Jesus Christ will come down from Heaven to dwell eternally with the children of God. The Father and Jesus sit on their thrones in Heaven now, where they dwell until Jesus creates the new heavens and Earth. Then they come down to the new Earth.

> I heard a loud shout from the throne, saying, "Look, God's home is now among his people! He will live with them, and they will be his people. God Himself will be with them." (Revelation 21:3 NLT)

The following compares the creation of the original Earth with the new Earth. This compares Genesis chapter 1 with Revelation chapters 21 and 22.

| Day | Original Creation | New Earth Creation |
|---|---|---|
| 1 | Light; day and night | Jesus is light of the new Earth; No night and darkness |
| 2 | Sky (atmosphere around Earth) | Possibly no sky; no separation of waters |
| 3 | Seas, dry ground, and vegetation | No seas; entire Earth is dry ground; Possibly vegetation |
| 4 | Sun, moon, stars; seasons, years, and days | Stars, no solar system for Earth; Jesus is the light that shines on Earth; Time not needed |
| 5 | Sea life and birds of the air | No sea life since no seas; Possibly bird life |
| 6 | Domestic & wild land creatures; People | Possibly land animals (pets?); No reproduction; Only believers there |
| 7 | Creation finished (Day of rest) | New Earth is place of eternal rest because no curse exists |

Table 3: Comparison of original creation with creation of new Earth

Note: The "sky" is the atmosphere around the Earth where birds fly. There doesn't seem to be a need for it around the new Earth.

## WHAT THE NEW EARTH WILL BE LIKE

In this section, we review scriptures that describe what the new Earth will be like. We can make assumptions about whether and how they might exist from these.

### Jesus Provides Its Light

It's impossible for people today to see the glory of the divine God and live (Exodus 33:20). Yet, in our resurrected eternal bodies, we will see the full glory of the Father and Jesus (he illuminates the entire Earth).

> And the city has no need of sun or moon, for the glory of God illuminates the city, and the Lamb is its light. The nations will walk in its light, . . . Its gates will never be closed at the end of day because there is no night there. (Revelation 21:23–25 NLT); (see also Isaiah 60:19)

Note: Glory is the brilliant brightness of God's supernatural divine presence.

Instructional Comment: We read about the light of Jesus in the New Testament. It does not refer to the physical light of his glory that will be seen on the new Earth. It is symbolic. This light (or enlightenment) refers to the truth he brings that illuminates people's hearts and minds. It's a spiritual enlightenment of the soul.

> Again Jesus spoke to them, saying, "I am the light of the world. Whoever follows me will not walk in darkness, but will have the light of life." (John 8:12 ESV)

### No Oceans or Seas

There will be no oceans or seas on the new Earth. There apparently will be no need for these to supply oxygen, food, or an avenue for transportation (sea-going vessels).

> Then I saw a new heaven and a new earth, for the old heaven and the old earth had disappeared. And the sea was also gone. (Revelation 21:1 NLT)

Instructional Comment: The surface of our current Earth is comprised of three-fourths seas and oceans. They provide one-half of the oxygen on Earth (some scientists say 70 percent). The remaining oxygen comes from trees and plant life (especially rainforests).

### No Freshwater (Other than River of Life)?

The only freshwater mentioned in Revelation chapters 21 and 22 is the River of Life that flows from the throne of God in the New Jerusalem. Does this imply there will be no freshwater for drinking and farming throughout the new Earth?

> Then the angel showed me a river with the water of life, clear as crystal, flowing from the throne of God and of the Lamb. It flowed down the center of the main street. (Revelation 22:1–2 NLT)

### Animals on the New Earth?

Since docile and wild animals existed in God's perfect creation of the original Earth (Genesis 1:20–25), should we expect this to be true on the new Earth? I suspect people and animals will dwell safely together eternally on the new Earth. Since people will have eternal resurrected bodies with no marriage (Matthew 22:30) or reproduction, I believe animals will be eternal and will not reproduce offspring.

### Figurative or Literal Language about Animals on the New Earth?

The following prophetic verses may be confusing. Are they literal about the new Earth or figurative illustrating the Father God's love, blessings, and protection of his people, Israel?

> The wolf shall dwell with the lamb, and the leopard shall lie down with the young goat, and the calf and the lion and the fattened calf together; and *a little child shall lead them*. The cow and the bear shall graze; their young shall lie down together; and the lion shall eat straw like the ox. The *nursing child* shall play over the hole of

the cobra, and the *weaned child* shall put his hand on the adder's den. They shall not hurt or destroy in all my *holy mountain*; for the earth shall be full of the knowledge of the LORD as the waters cover the sea. (Isaiah 11:6–9 ESV, author's emphasis)

Instructional Comment: There is no indication in the Bible that babies and little children will be on the new Earth. So, are they figurative or literal in these verses? We will have to wait and see.

## New Jerusalem's Gates Never Close

The following prophecy says the gates of the city of the New Jerusalem will never be closed. Therefore, we can conclude there will be no fear of anything harming God's people on the new Earth. If there are animals there, they will not be dangerous.

> By its light will the nations walk, and the kings of the earth will bring their glory into it, and its gates will never be shut by day—and there will be no night there. (Revelation 21:24–25 ESV)

## Pets

Some people have beloved pets and wonder if they will be with them in Heaven and the new Earth. I do not find anything in the Bible to answer this heartfelt question. I personally hope there will be.

## CHAPTER QUESTIONS

DQ1: What is God's purpose for creating the new Earth?

DQ2: Why must it be a uniquely new creation and not a renovation of the current corrupted Earth?

DQ3: Why do you think animal will be on the new Earth?

DQ4: Why will the gates of the city of the new Jerusalem never need to be closed?

## DIG DEEP

DD1: Why will the new Earth and universe be created out of nothing? Why will Jesus use a new eternal Periodic Table to do so?

## PERSONAL APPLICATION

P1: What promises of the Father God about the new Earth do you look forward to experiencing?

# 38.

## Creation of a New Earth: People

The Father's eternal provisions for his people on the new Earth will be beyond our imagination. In this chapter, we will examine what believers will experience on the new Earth.

### IS ISAIAH 65:17-25 FIGURATIVE LANGUAGE?

God sometimes uses figurative language in the Bible to express his blessings, promises, and benefits for his people (usually Israel). This is particularly true in the Old Testament. They are not intended to be understood and applied in a literal manner.

Isaiah 65:17-25 may be such figurative language. The verses seem to describe a time of peace and blessings from the Father God for his children, Israel, on this Earth (not the new Earth). The promised blessings will be so wonderful that Israel will feel like they have entered the future, literal new heavens and new Earth. Please read Albert Barnes' notes after these verses.

Following are these non-literal figures of speech in Isaiah 65:17-25 that illustrate the Father's love for his children, the nation of Israel on this Earth:

> Look! I am creating new heavens and a new earth, and no one will even think about the old ones anymore. Be glad; rejoice forever in My creation! And look! I will create Jerusalem as a place of happiness. Her people will be a source of joy. I will rejoice over Jerusalem and delight in My people. And the sound of weeping and crying will be heard in it no more. No longer will babies die when only a few days old. No longer will adults die before they have lived a full life. No longer will people be considered old at one hundred! *Only the cursed will die that young! In those days people will live in the houses they build and eat the fruit of their own vineyards.* Unlike the past, invaders will not take their houses and confiscate their vineyards. For My people will live as long as trees, and *My chosen ones will have time to enjoy their hard-won gains.* They will not work in vain, and their children will not be doomed to misfortune. For they are people blessed by the LORD, and their children, too, will be blessed. I will answer them before they even call to Me. While they are still talking about their needs, I will go ahead and answer their prayers! The wolf and

the lamb will feed together. The lion will eat hay like a cow. But the snakes will eat dust. In those days no one will be hurt or destroyed on My holy mountain. I, the LORD, have spoken! (Isaiah 65:17–25 NLT, author's emphasis)

Abert Barnes[1] describes these verses as poetry as follows:

*The passage before us is highly poetical, and we are not required to understand it literally . . . ; and all that the language necessarily implies is, that there would be changes in the condition of the people of God as great as if the heavens, overcast with clouds and subject to storms, should be recreated, so as to become always mild and serene; or as if the earth, so barren in many places, should become universally fertile and beautiful. The immediate reference here is, doubtless, to the land of Palestine, and to the important changes which would be produced there on the return of the exiles*; but it cannot be doubted that, under this imagery, there was couched a reference to far more important changes and blessings in future times under the Messiah *(literal new heavens and new Earth of Revelation chapters 21 and 22)*—changes as great as if a barren and sterile world should become universally beautiful and fertile. (Albert Barnes Notes on the Bible, author's emphasis)

## RESURRECTED PHYSICAL BODY

How will believers' resurrected, eternal bodies be different from their current physical bodies? I believe, for one thing, it will be composed of eternal elements like that of Jesus.

So also is the resurrection of the dead. It is sown a perishable body, it is raised an imperishable body. (1 Corinthians 15:42 NASB)

I believe we will be able to do the same things Jesus did on Earth in his resurrected body.

When the doors were shut where the disciples were, for fear of the Jews, Jesus came and stood in their midst and said to them, "Peace be with you." (John 20:19 NASB)

When He *(Jesus)* had reclined at the table with them, He took the bread and blessed it, and breaking it, He began giving it to them. Then their eyes were opened and they recognized Him; and He vanished from their sight. (Luke 24:30–31 NASB, author's emphasis)

### Babies and Young Children

I have read about people who had visions of Heaven and saw babies and young children there. They wonder if this means that these will remain babies and young children eternally on the new Earth. I suspect no babies, children, or older adults will be on the new Earth.

Instructional Comment: I don't see any scriptures that directly answer these questions. From the book of Revelation, it appears every person on the new Earth (animals and plants, if they are there) will be fully grown and mature. If true, God might miraculously

---

1. Barnes, *Notes*, Isaiah 65:17–25.

and instantly transform babies and children to a mature age in Heaven to prepare them for the new Earth. Just conjecture.

## No Old Nature

Since believers will be like their Savior, Jesus Christ, on the new Earth, they will no longer have an old, self-centered, sin-based nature with which to battle.

> For I know that nothing good lives in me, that is, in my flesh. For I want to do the good, but I cannot do it. For I do not do the good I want, but I do the very evil I do not want! (Romans 7:18–19 NET)

They will have the perfect nature of their Savior and Lord, Jesus Christ. I believe they will rejoice on the new Earth because they no longer have to wage war with their old self.

> For if we have become united with Him in the likeness of His death, certainly we shall also be in the likeness of His resurrection. (Romans 6:5 NASB)

## Fictional Instant Transportation

The movie Star Trek introduced a fictional way for people to have their physical bodies transported from the starship to a planet's surface. The fanciful process of beaming a person involves dematerializing them at one location and rematerializing them at a new location. Perhaps this idea may not be so far from the truth on the new Earth. Just as Jesus instantly appeared in different locations after his resurrection, perhaps believers may be able to do the same.

# GOD'S LOVE ON THE NEW EARTH

Born-again believers receive the Holy Spirit and God's love in their hearts on this Earth. However, their old nature limits their ability to constantly love God and one another with his eternal, unconditional love. Even at best, their love is conditional, often based on what they can get from it.

This is made clear in the following Scripture:

> If I speak in the tongues of men and of angels, but have not love, I am a noisy gong or a clanging cymbal. And if I have prophetic powers, and understand all mysteries and all knowledge, and if I have all faith, so as to remove mountains, but have not love, I am nothing. If I give away all I have, and if I deliver up my body to be burned, but have not love, I gain nothing. Love is patient and kind; love does not envy or boast; it is not arrogant or rude. It does not insist on its own way; it is not irritable or resentful; it does not rejoice at wrongdoing, but rejoices with the truth. Love bears all things, believes all things, hopes all things, endures all things. Love never ends. As for prophecies, they will pass away; as for tongues, they will cease; as

for knowledge, it will pass away. For we know in part and we prophesy in part, but when the perfect comes, the partial will pass away. (1 Corinthians 13:1–10 ESV)

Since believers will no longer have their self-centered old nature, I believe they will constantly experience and share God's eternal love in their hearts without the conflicting old nature.

---

Believers on the new Earth will constantly experience
God's eternal love in their hearts and lives.

---

### God's Love, His Name Written on Believers

God is love by his nature (1 John 4:8). His name represents who he is, his nature. Therefore, when he writes his name on believers on the new Earth, he identifies them as belonging to him and recipients of his eternal love.

The following verse says the Father and Jesus will supernaturally write their names to believers on the new Earth. God will also write the name of the new Jerusalem on them.

> The one who conquers, I will make him a pillar in the temple of my God. Never shall he go out of it, and I will write on him the name of my God, and the name of the city of my God, the new Jerusalem, which comes down from my God out of heaven, and my own new name. (Revelation 3:12 ESV)

This may be figurative language and not literal writing on the bodies of believers on the new Earth. Perhaps it refers to the symbolic writing on the hearts and minds of the people of God, as we see in the following verses about the New Covenant:

> For this is the covenant that I will make with the house of Israel after those days, declares the Lord: I will put my laws into their minds, and write them on their hearts, and I will be their God, and they shall be my people. And they shall not teach, each one his neighbor and each one his brother, saying, "Know the Lord," for they shall all know me, from the least of them to the greatest. (Hebrews 8:10–11 ESV)

## CITIES AND PEOPLE

Many believers will live on the new Earth, perhaps hundreds of millions. Therefore, they will likely be spread over the new Earth, though some will live in the city of the New Jerusalem.

> The nations will walk by its light, and the kings of the earth will bring their glory into it. (Revelation 21:24 NASB)

*Nations*, Greek is *ethnos*; meaning people, race, tribe, gentiles.

Who are these nations? Perhaps the best English word is "people." Maybe this means that people from all the nations of the world, including Jews and Gentiles, will inhabit the entire new Earth.

## Leaders and Government

Will there be a need for governing authorities in these cities around the new Earth? If so, who is ruling over them? Will they be Christians who were successful politicians, business leaders, pastors, or believers with spiritual leadership gifts?

> The kings of the earth will bring their glory into it. (Revelation 21:24 NASB)

> And he said to him, "Well done, good slave, because you have been faithful in a very little thing, you are to be in authority over ten cities...And he said to him also, 'And you are to be over five cities." (Luke 19:17–19 NASB)

Instructional Comment: Perhaps these verses imply God will create a form of government on the new Earth. I am not certain of its value since there will be no crime or need to punish or reward human behavior as there is today (Romans 13:1–5).

## Jesus Will Be the Government

In the following Scripture, we see that Jesus will govern his people. Does this imply that he alone will govern the cities on the new Earth? Or do these verses simply refer to his eternal sovereignty over all creation as King of kings and Lord of lords (1 Timothy 6:15)?

> For a child will be born to us, a son will be given to us; And the government will rest on His shoulders; And His name will be called Wonderful Counselor, Mighty God, Eternal Father, Prince of Peace. (Isaiah 9:6 NASB)

# NO MARRIAGE

Jesus states that people will not be married in Heaven. Believers will be like the angels in Heaven regarding marriage. Their physical needs and impulses will no longer govern (drive) them. I assume this also applies to the new Earth.

> For in the resurrection they neither marry nor are given in marriage, but are like angels in heaven. (Matthew 22:30 NASB)

Instructional Comment: There are many things that the new Earth has in common with Heaven. I suspect the above verse applies to no marriage and no reproduction on the new Earth.

## NO NEED FOR FOOD

It appears from the following New Testament verse that people may not need to eat food on the new Earth.

> Food is for the stomach and the stomach is for food, but God will do away with both of them. (1 Corinthians 6:13 NASB)
>
> *Do Away with*, Greek is *katargeō*; to render inoperative, abolish.

Instructional Comment: Perhaps believers will not have a physical need to eat to nourish and sustain their eternal bodies on the new Earth. Jesus ate in his resurrected body before his ascension into Heaven. I don't think he needed to eat to nourish his eternal body.

### Plant Life

Since there was plant life in God's original perfect creation (Genesis 1:11), I suspect there may be beautiful eternal plant life on the new Earth. Perhaps some of it will be edible.

### Naked or Clothes?

An interesting thought is whether believers will wear clothes on the new Earth. Adam and Eve did not have clothes; they were naked and were not embarrassed by that (Genesis 3:10). If believers do wear clothes on the new Earth, what will they be made of: cotton, wool? If so, there must be farms to raise the cotton and the sheep for the wool. This means work for some as farmers.

Instructional Comment: Adam and Eve did not feel shame or embarrassment when naked in the Garden of Eden before rebelling against God. If people are naked on the new Earth, I suspect it will be the same.

## WORK?

Adam and Eve had jobs as farmers in the Garden of Eden (Genesis 2:15), but I am unsure if that implies believers will have jobs on the new Earth.

### If There Is Work, It Will Not Be Toil

God caused work after to be toilsome and laborious after the fall.

> Then to Adam He said, "Because you have listened to the voice of your wife, and have eaten from the tree about which I commanded you, saying, 'You shall not eat from it'; cursed is the ground because of you; in toil you will eat of it all the days of your life. Both thorns and thistles it shall grow for you; and you will eat the plants of the field; by the sweat of your face You will eat bread, till you return to the ground." (Genesis 3:17–19 NASB)

### If There Is Work, It Will Be Free from Sin

If there is work on the new Earth, it will be an environment free from the effects of the fall and curse. There will be no physical and emotional pain, frustration, stress, corrupt or unreasonable bosses, interpersonal conflicts, racial discrimination, greed, and competition to be the best and get to the top of a career ladder.

> He will wipe away every tear from their eyes, and death shall be no more, neither shall there be mourning, nor crying, nor pain anymore, for the former things have passed away. And he who was seated on the throne said, "Behold, I am making all things new." Also he said, "Write this down, for these words are trustworthy and true." (Revelation 21:4-5 ESV)

### Rest from Toil on New Earth

The following verses tell us that we will have rest from our toils on the new Earth.

> For we who have believed enter that rest, . . . For He has said somewhere concerning the seventh day: and God rested on the seventh day from all his works; . . . So there remains a Sabbath rest for the people of God. For the one who has entered His rest has himself also rested from his works, as God did from His. (Hebrews 4:3-10 NASB)

### People Will Live in Homes

Jesus promised his followers there would be places for them to live in Heaven. We can assume this also means on the new Earth.

> There are many dwelling places in my Father's house. Otherwise, I would have told you, because I am going away to make ready a place for you. And if I go and make ready a place for you, I will come again and take you to be with me, so that where I am you may be too. (John 14:2-3 NET)

> *House*, Greek is *oikia*; residence, an abode, house.

> *Dwelling Places, Rooms*, Greek is *monē*; staying, that is, residence.

## TREE OF LIFE

The Tree of Life is one of the greatest mysteries of the new Earth.

> On either side of the river was the tree of life, bearing twelve kinds of fruit, yielding its fruit every month. (Revelation 22:2 NASB)

> Blessed are those who wash their robes, so that they may have the right to the tree of life, and may enter by the gates into the city. (Revelation 22:14 NASB)

Besides its fruit, another Mystery is why people will use the leaves from the Tree of Life.

> And the leaves of the tree are for the healing of the nations. Revelation 22:2 (NASB); (also a prophecy of Tree of Life, Ezekiel 47:12)
>
> *Heal*, Greek is *therapeia*; (medical, cure); attend to, healing.

John Gill[2] says the following about the leaves of the Tree of Life:

> These leaves will be for the preserving and continuing the health of the people of God in this state, as the Tree of Life in Eden's garden was for the preservation of the health and life of Adam, had he continued in a state of innocence. (John Gill's Expository of the Entire Bible)
>
> Mystery: What does eating the fruit of the Tree of Life represent? Why is there a need for believers to use the leaves of the Tree of Life for healing since they already have eternal life?

## WATER OF LIFE

The Water of Life is a river that flows from the thrones of God the Father and Jesus in the New Jerusalem on the new Earth. Notice that it's not a weak, small stream of water. It is significant and has supernatural abilities and effects. We cannot guess what the Water of Life is and how it benefits people on the new Earth. But it will be a rich and wonderful blessing to everyone there.

> Then the angel showed me a river with the water of life, clear as crystal, flowing from the throne of God and of the Lamb. (Revelation 22:1–2 NLT)
>
> And He also said, "It is finished! I am the Alpha and the Omega—the Beginning and the End. To all who are thirsty I will give freely from the springs of the water of life." (Revelation 21:6 NLT)

## MANY MYSTERIES OF THE NEW EARTH

There are many mysteries about the new heavens and new Earth that we cannot begin to imagine what they might be. The following Bible verse tells us this:

> That is what the Scriptures mean when they say, "No eye has seen, no ear has heard, and no mind has imagined what God has prepared for those who love Him." (1 Corinthians 2:9 NLT)

How can you know if you are destined for the new Earth? Why should it matter to you now? Don't you have plenty of time to decide? God says now is the time for salvation (2 Corinthians 6:2).

---

2. Gill, *Expository*, Revelation 22:2.

## CHAPTER QUESTIONS

DQ1: Why do you think Isaiah 65:17–25 is figurative language that should not be understood literally about the new universe and Earth of Revelation chapters 21 and 22?

DQ2: Why must every believer on the new Earth have an eternal resurrected body like that of Jesus Christ?

DQ3: Since angels spend their time worshipping and serving God, what will believers be doing for eternity on the new Earth?

DQ4: How is the Tree of Life on the new Earth different from evolution's biological Tree of Life theory?

## DIG DEEP

DD1: There are mysteries concerning the purpose of the leaves of the Tree of Life. What do you think their purpose is if there is no sickness or death on the new Earth?

## PERSONAL APPLICATION

PA1: Why won't you get bored worshipping and serving the Father God and his Son, Jesus Christ, on the new Earth?

# PART L

Creation Is Truth

# 39.

# What Is Truth?

This is an opportunity to ask ourselves, "What is truth?" Which is true, creation or the theory of evolution? Pontius Pilate mocked Jesus when he asked him what truth was (just before he crucified him).

> "I was born and came into the world to testify to the truth. All who love the truth recognize that what I say is true." "What is truth?" Pilate asked. (John 18:37–38 NLT)

## JESUS DEFINED TRUTH

Only the God of the Bible can define truth. Jesus does so in the following verses:

> Jesus told him, "I am the way, the truth, and the life. No one can come to the Father except through Me." (John 14:6 NLT)
>
> And you will know the truth, and the truth will set you free. (John 8:32 NLT)
>
> But the time is coming—indeed it's here now—when true worshipers will worship the Father in spirit and in truth. The Father is looking for those who will worship Him that way. (John 4:23 NLT)

Truth is the reality of the way things are. It's the nature of things and not theories about them. The Trinity is truth. Jesus is truth. The Father is truth. The Holy Spirit is the "Spirit of Truth" (John 15:26).

## CREATION IS TRUTH

We have reviewed many things that Christians know to be true about creation.

The following are three essential things Christians know about the creation account that are true:

- There is an Intelligent Designer behind everything created
- The Creator God (Trinity) is the author of creation
- The creation account in Genesis chapters 1 and 2 accurately describes what the Father designed and Jesus created [/NL 1–3]

## EVOLUTION IS NOT TRUTH

We have also reviewed many things that invalidate the various theories of evolution. Evolution is a theory and, therefore, not truth.

The following are four essential things that disprove the theory of the biological evolution of species:

- Evolutionary scientists are not able to rely on using DNA and the genetic composition of plants and animals alone to identify and classify species accurately.
- Neanderthal DNA and artifacts prove they are not the transitional species that evolved into modern humans (thus disproving the theory of human lineage from ancient ancestors).
- Fragile DNA extracted from viable dinosaur tissues proves they are thousands of years old, not millions of years old (DNA does not survive for millions of years).
- New genetic material can not be added to living organisms.

Important: Some evolutionists and scientists are beginning to recognize that the theory of evolution is false. Some even admit that there is an intelligent design to nature. Yet, they still deny creation and the existence of the Creator God.

## THEORIES ARE NOT TRUTH

If you read websites and articles written by evolutionists, you will see many still wholeheartedly believe in their theories. I have stated throughout this book that theories are not truth.

> Theories are not truth. At best, they are intelligent guesses. Creation is not a theory, it's truth.

In the following verse, God warns us not to believe an *empty, deceitful philosophy*:

> Be careful not to allow anyone to captivate you through an empty, deceitful philosophy that is according to human traditions and the elemental spirits of the world, and not according to Christ. (Colossians 2:8 NET)

## ETERNAL LIFE AND DEATH ARE AT STAKE

The following verse says that we are to be careful about what we believe because wrong thinking will always lead to (eternal) death:

> There is a way that seems right to a person, but its end is the way that leads to death. (Proverbs 14:12 NET)

What is this eternal death? In the Bible, it's referred to as the "second death." This is an eternal existence away from the love and presence of God in the Lake of Fire.

> This lake of fire is the second death. And anyone whose name was not found recorded in the Book of Life was thrown into the lake of fire. (Revelation 20:14–15 NLT)

In the next verses, Jesus says that many people will choose to ignore the truth:

> Enter through the narrow gate. For wide is the gate and broad is the road that leads to destruction, and many enter through it. But small is the gate and narrow the road that leads to life, and only a few find it. (Matthew 7:13–14 NIV)

## MYSTERIES, FACTS, THEORIES, AND MYTHS

I have presented many mysteries, facts, theories, and myths for your consideration. How have your ideas and beliefs about God, the Bible, creation, and the theory of evolution changed after reading this book? Were there any surprises for you?

## CHAPTER QUESTIONS

DQ1: Why can Jesus Christ be trusted to know what truth is?

DQ2: What are the three essential things we know about creation that are true?

DQ3: What are the four essential things we know that disprove the theory of evolution?

DQ4: Why are there so many mysteries beyond human imagination in the Bible and in life?

## DIG DEEP

DD1: Jesus said, "I am the way, and the truth, and the life; no one comes to the Father but through Me" (John 14:6 NASB). Jesus is truth in his nature (as is the Father and Holy Spirit). This means he will never lie. What does his statement say about the possibility of eternal life with God through other religions? Why is his statement both inclusive and exclusive?

## PERSONAL APPLICATION

PA1: Why do you know you can trust the Father God, Jesus Christ, and the Holy Spirit?

# 40.

## Conclusion: It's Up to You!

### EVERYONE BELIEVES IN SOMEONE OR SOMETHING

Do you believe in the Creator God of the Bible? People put their faith in what they believe to be true. The perception of truth is the basis on which beliefs are built. Yet, perceptions can be false. Where do you go to find truth? We should seek truth in the Bible because truth is the substance of reality revealed by the God of the Bible. Remember that theories about the Bible are not truth.

### Dead Dinosaurs and Dead Neanderthals Don't Lie

Studying the sciences can reveal truth. But theories about science are not truth. Remember that dead dinosaurs and dead Neanderthals don't lie. Why? Because science and their DNA prove they existed thousands of years ago, not millions of years ago, according to theories from evolution.

### Does the God of the Bible Exist?

Apparently, not to 96 percent of adult Americans. George Barna's 2023 Biblical Worldview survey concludes that 96 percent of adult Americans have created their own idea of who they want God to be. That means only 4 percent view the world through the lens of Scripture. It means that only 4 percent believe in and live for God as he describes himself in the Bible. Many studies indicate unbelief in the God of the Bible exists worldwide.

---

Worldview: Does the God of the Bible really exist? How can you know if he does? He says you can know by reading the Bible. You will see that he wants you to know him personally.

---

Website: Information and details about George Barna's 2023 Biblical Worldview survey can be found on the following website: https://www.arizonachristian.edu/wpcontent/uploads/2023/02/CRC_AWVI2023_Release1.pdf

## WHERE YOU CAN FIND TRUTH

Each person has to decide what is true and false. Do you believe the Bible is the word of the God of the Bible? Do you believe the entire Bible, from Genesis to Revelation, is accurate and true?

God says in the following verses that he inspired the entire Bible, and all of it is trustworthy and true:

> But you must remain faithful to the things you have been taught. You know they are true, for you know you can trust those who taught you. You have been taught the holy Scriptures from childhood, and they have given you the wisdom to receive the salvation that comes by trusting in Christ Jesus. *All Scripture is inspired by God and is useful to teach us what is true and to make us realize what is wrong in our lives. It corrects us when we are wrong and teaches us to do what is right.* (2 Timothy 3:14–16 NLT, author's emphasis)
>
> *Since all Scripture in the Bible is from God and is true, the account of creation is true.*

## WHAT KIND OF FAITH?

There are two kinds of faith in this world: Blind faith based on assumptions and unproven theories and informed faith based on truth and reality.

### Evolution Is Blind Faith

What kind of faith is required to believe that living organisms evolved from lifeless chemicals (the theory of biological evolution)? And where did these chemicals come from? Did they create themselves? Evolution is the classic story of "blind faith," a faith with no reality or substance.

### Creation Is Informed Faith

The Christian faith in the God of the Bible and creation is not blind faith. You have read about many archaeological discoveries that have proven the Bible to be true. You have also seen how many sciences undergird the Bible and the biblical creation account. Christianity is an informed faith. It is a substantiated faith.

## CHRISTIAN'S RESPONSE TO EVOLUTIONISTS

How should Christians respond to people, whether scientists or lay people, who believe in and support the theory of evolution? We should treat them with dignity and respect as people loved by the Creator God. We do not accept their beliefs as accurate, but we can have civil conversations and develop relationships with them. Perhaps the truth of the Spirit in you will illuminate a path for their salvation.

> Correcting opponents with gentleness. Perhaps God will grant them repentance and then knowledge of the truth and they will come to their senses and escape the devil's trap where they are held captive to do his will. (2 Timothy 2:25–26 NET)

The next verse says that you must strive to live in peace with all people as far as it is up to you:

> If possible, so far as it depends on you, live peaceably with all people. (Romans 12:18 NET)

## GOOD NEWS FOR EVERYONE

The Good News (gospel) about Jesus Christ will save anyone who puts their faith in him as Savior and Lord.

> For I am not ashamed of this Good News about Christ. It is the power of God at work, saving everyone who believes—the Jew first and also the Gentile. This Good News tells us how God makes us right in His sight. This is accomplished from start to finish by faith. As the Scriptures say, "It is through faith that a righteous person has life." (Romans 1:16–17 NLT)

## IN THE END, IT'S UP TO YOU!

God wants everyone to spend eternity with him, no matter who they are and what they have believed. The heart of the gospel message is the free gift of salvation through Jesus Christ.

> Who desires all people to be saved and to come to the knowledge of the truth. (1 Timothy 2:4 ESV)

> For "everyone who calls on the name of the Lord will be saved." (Romans 10:13 ESV)

### Do You Want to Know God Better?

If you want a book to help you know the God of the Bible better, I suggest our first book, *Who Is This God? A Handbook for Life with Him*. It can be purchased at major online stores that sell books (Amazon, Barnes and Noble, and more).

### Jesus Invites You into His Eternal Kingdom

You have studied the Bible from its first verses in Genesis to its last verses in Revelation. You have learned many truths that can set you free from false beliefs (John 8:32).

Following are the final words of Jesus Christ in the Bible:

> "I, Jesus, have sent My angel to give you this message for the churches." . . . The Spirit and the bride say, "Come." Let anyone who hears this say, "Come." Let anyone who is thirsty come. Let anyone who desires drink freely from the water of life . . . . He who is the faithful witness to all these things says, "Yes, I am coming soon!" Amen! *Come, Lord Jesus!* (Revelation 22:16–20 NLT, author's emphasis)
>
> *"Come, Lord Jesus!"*

## CHAPTER QUESTIONS

DQ1: Why do you think everyone believes in someone or something? What are some examples of these beliefs that do not pertain to the Bible or evolution?

DQ2: Why is creation an informed faith and evolution a blind faith?

DQ3: Why should Christians be gracious and kind when challenged by evolutionists about their faith in the God of the Bible and creation?

DQ4: Why does God want everyone to spend eternity with him?

## DIG DEEP

DD1: What does the statement, "Everyone believes in someone or something," have to do with a person's worldview?

## PERSONAL APPLICATION

PA1: If you have not asked Jesus Christ to be your Savior and Lord, then "today is the day of salvation" for you. Please read the chapter, "Prayer to Know God" to discover how to enter into an eternal relationship with the Creator God of the Bible.

# Prayer to Know God

Sometimes, people ask, "Isn't being a good person enough to get into Heaven?" But we must ask ourselves the following questions: "How good is good enough?" and "Whose measuring stick for goodness are we using?"

Have you ever done anything wrong in your life? If so, you are not perfect and not good enough to enter Heaven.

> Your eyes are too pure to look on evil; you cannot tolerate wrongdoing. (Habakkuk 1:13 NIV)
>
> You therefore must be perfect, as your heavenly Father is perfect. (Matthew 5:48 ESV)

## No One Can Earn Salvation

No one is perfect enough to live in God's holy presence. Every person (no matter how nice they are) has said, thought, and done things that displease God. He calls these behaviors of wrongdoing "sin." The Greek word translated *sin* is sometimes referred to as "missing the mark" of God's perfection.

> For everyone has sinned; we all fall short of God's glorious standard. (Romans 3:23 NLT)

What about the person who says, "I have always been a Christian." Can this be true? No, it's not true. Instead, the opposite is true.

> And remember that those who do not have the Spirit of Christ living in them do not belong to Him at all. (Romans 8:9 NLT)

The Holy Spirit dwells within people only when they commit their hearts and lives to Jesus as their Savior and Lord.

### There Is No Hope without Jesus Christ

The following verses describe a person's spiritual condition without Jesus Christ.

> In those days you were living apart from Christ. You were excluded from citizenship among the people of Israel, and you did not know the covenant promises God had made to them. You lived in this world without God and without hope. (Ephesians 2:12 NLT)

> For the sinful nature is always hostile to God. It never did obey God's laws, and it never will. That's why those who are still under the control of their sinful nature can never please God. (Romans 8:7–8 NLT)

I heard a pastor say that you "must step across the line" to intentionally choose to leave your old, condemned life and accept God's new eternal life in Christ. It's the most important decision you will ever make.

> There is no judgment against anyone who believes in Him *(Jesus)*. But anyone who does not believe in Him has already been judged for not believing in God's one and only Son. (John 3:18 NLT, author's emphasis)

Some born-again followers of Jesus cannot state a specific date or event when they made this decision. But they know in their hearts that they have committed themselves to him as their Savior and Lord.

### There is Good News!

It's God's provision to rescue you from the eternal consequences of living your life without Jesus. You can be saved from these consequences and gain an eternal relationship with the Father God through Jesus Christ.

## YOUR PRAYER TO KNOW GOD

The following prayer is the Good News. Please pray this from your heart to enter an eternal relationship with the Father, Jesus, and the Holy Spirit.

> Dear God, thank you for loving me and caring about me. Thank you for wanting me to be with you forever and for sending Jesus to be my Savior and Lord. I believe that Jesus died on the cross for me, that he was dead and buried, and was raised back to life by your supernatural power, according to the scriptures.

> Please forgive me for all my sins against you and people. I commit to live my life under your authority and rule, Jesus, as the Lord of my life. Thank you for rescuing me from the eternal consequences of living without you.

> Thank you, Father, for now sending your Holy Spirit to live within me to guide, help, and teach me. I know that I have eternal life with you, my God!

> *If you sincerely prayed the above from your heart, you are now a born-again Christian, a follower of Jesus Christ.* The Holy Spirit is living within you. Welcome to the eternal family of God!

You now have the following relationship with each person of the Trinity of God. You are a:

- Child of your Father God (Galatians 3:26)
- Disciple and brother/sister of Jesus Christ (Mark 3:35)
- Temple in which the Holy Spirit lives (1 Corinthians 6:19)

## CHANGES AFTER BEING BORN-AGAIN

When you are born-again, many wonderful changes occur for you.

The following is a list of some changes that occur after being born-again. They describe what you receive as a new follower of Jesus Christ.

- You are a chosen child of the Father God. (2 Thessalonians 2:13; Ephesians 1:4)
- You are rescued and forgiven by God; you are made right with God. (Romans. 5:9; Colossians 1:14)
- You are transferred from the devil's domain to Christ's Kingdom. (Colossians 1:13)
- You are born-again because the Holy Spirit came to live within you. (1 Corinthians 3:16; Ephesians 2:22)
- The Father God and Jesus came with the Holy Spirit to live within you. (John 14:23)
- The Holy Spirit within you is the Father's guarantee he loves you and wants you to be with him eternally. (2 Corinthians 1:21–22)
- You will always belong to God. (Romans 14:7–9)
- The Father God adopted you as his child. (John 1:12; Ephesians 1:5)
- You can never be separated from God's love. (Romans 8:35)
- You can never be condemned for your forgiven sins. (Romans 8:1,2, 31)
- Jesus Christ calls you his friend. (John 15:15)
- You are one in Spirit with Jesus Christ. (1 Corinthians 6:17)
- You are a member of the eternal body of Jesus Christ. (1 Corinthians 12:27)
- You have a new nature that is continually being transformed to be like Jesus. (2 Corinthians 5:17; Galatians 6:15)
- God calls you a "saint" because you are now blameless in his sight. You have received the perfection of Jesus. (Ephesians 1:1)
- You have been given spiritual gifts to serve other people. (1 Corinthians 12:7; 1 Peter 4:10)
- God works in all things for the good of those who love him. (Romans 8:28)
- God is at work within you to continually transform you to be more like Jesus Christ. He will do so until you are home in Heaven with him. (Philippians 1:6)

- Your eternal citizenship is now in Heaven. (Philippians 3:20)
- You have not been given a spirit of fear but of power, love, and a sound mind. (2 Timothy 1:7)
- You may approach your Father God in prayer with confidence, knowing he always wants you to do so. (Ephesians 2:18, 3:12; Hebrews 4:16)
- Jesus Christ gives you his strength to do what he asks you to do. (Philippians 4:13)
- You are a witness to others about what Jesus has done for you. (John 15:26–27; Acts 1:8; 2 Corinthians 5:17)

**Communing with the God of the Bible**

The Holy Spirit will enable you to commune and communicate with each person of the God of the Bible. It is wonderful to have an eternal, living relationship with the Father, Jesus, and the Holy Spirit as a new follower of Christ! *Hallelujah, and welcome home!*

# Glossary

The following alphabetical glossary provides the meaning of key terms in this book.

Aliens: Aliens are theorized to be intelligent, sentient beings who exist on other planets and galaxies. Extraterrestrial life is another name used for aliens, meaning life that might exist outside the Earth and its atmosphere.

Allele: One of two or more variations of a gene on the same place on a chromosome. Each parent provides one version of the allele. Different alleles can result in different traits.

Archaeology: This science studies the material evidence of ancient and recent human cultures. There are three main types of archaeology: (1) Prehistoric archaeology: Study of human cultures that did not have writing; (2) Protohistoric archaeology: Study of human cultures that have incomplete records; (3) Historic archaeology: Study of human cultures that have well-developed historical records. Historical records could be written or oral in nature. Artifacts (tools, weapons, pottery, and such) from these three types provide archeologists with information about the people, their society, religious beliefs, and culture.

Biology's Tree of Life Theory: This is also called a "phylogenetic tree." It depicts the theoretical evolutionary relationships for organisms as they evolved from their beginning into current species. In this theory, each upward branch on the tree is a new species that evolved over millions or billions of years from a previous species.

Bipedalism: A species classified as bipedal walks on two legs.

Book of Life: The Book of Life contains the names of everyone who accepts Jesus Christ as Savior and Lord. Some believe that it initially contained the names of everyone who ever lived, but their names were erased if they did not accept Jesus Christ. Others believe it initially contained only the names of those who would accept Jesus Christ.

Chromosome: Thread-like strings of DNA tightly coiled around proteins inside the nucleus of cells. People have two pairs of 23 chromosomes, one pair from each parent, for a total of 46.

Cladograms: These illustrate a hypothetical relationship between species based on physical traits in order to trace the species back to a common ancient ancestor. They are used by many evolutionists instead of direct descent lineage for species identification and classification.

Conscience: People have a conscience as part of the image of God (it's part of the human soul). It's a warning system concerning personal violations of right and wrong conduct. An inner sense (self-awareness) judges a person's actions, words, and motivations. Right and wrongs are moral standards, a set of moral codes of conduct unconsciously absorbed or consciously adopted.

Cosmology: The study of the universe and how it functions.

Cosmos: Another name for the universe.

Creation: The Bible describes creation in the first two chapters of the Old Testament book of Genesis. The first verse in the Bible says that the Triune God existed before creation, *In the beginning, God created the heavens and the earth* (Genesis 1:1 ESV).

Creation Kind(s): They were created by Jesus and would be classified at the species level, not the genus level. I see these kinds as "master species" that carry all the genetic material needed for generations of reproduction, leading to a variety of species as they reproduce according to the command of Jesus.

Creationism: The belief that the God of the Bible created everything according to Genesis chapters 1 and 2.

Creationists: People here who believe the biblical truths about creation but don't necessarily support all of the Young and Old Earth Creationists. Young Earth Creationists: Believe the creation account in Genesis chapters 1 and 2 and that a creation day was 24 hours. Therefore, the age of the universe, Earth, and humankind is 6,000 years (some say up to 10,000 years). Old Earth Creationists: Believe the creation account in Genesis chapters 1 and 2 but do not support the idea that the universe, Earth, and humankind are 6,000 years old. They accept the possibility that the six days of creation were not literal 24-hour days but rather longer periods. They are not evolutionists.

Dark Energy: Cosmologists theorize the universe is expanding faster and faster. But what makes this happen if the universe is only composed of matter (including dark matter)? Well, again, there is nothing visible to explain this. Because of the evidence in space, they believe it is a form of invisible energy called "dark energy." They believe it comprises 68 percent of the universe.

Dark Matter: Scientists investigating the universe cannot explain why our universe, even planet Earth, remains intact. What makes them stay where they are in the universe? For example, why does the Earth remain in orbit around the sun? Why does our sun stay in orbit in our galaxy? To explain this unknown phenomenon, cosmologists theorize the existence of an invisible substance, "dark matter." Because it cannot be seen, they propose it exists by the observable evidence of an effect on planets, solar systems, galaxies, and other cosmic objects. That universal effect is gravity. They think it accounts for about 27 percent of the universe. This means that all solid matter, such as planets, suns, asteroids, etc., theoretically account for only 5 percent of the universe.

Darwinian Evolution (Theory of Evolution by Natural Selection): In 1859, Charles Darwin published his book, *On the Origin of Species*. His theory is known as the Theory of

Evolution by Natural Selection. He proposes that living organisms change over long periods as they adapt to their environment. Individuals that adapt well survive and produce offspring that will survive. Through natural selection, his theory says that diverse species could evolve from a common ancestor.

Devil and His Demons: They were once angels of God who rebelled and were cast out of Heaven. Now, they strive to bring hate, destruction, and evil into the world and the lives of people. They are sometimes referred to as evil spirits or the spiritual enemy of believers. Their eternal fate is the Lake of Fire, where they will never again be able to harass God's people.

Delving Deeper Chapters: These chapters include information that may be difficult to comprehend due to its complex nature. This is true of some information about creation, evolution, and science.

DNA: Most Deoxyribonucleic acid (DNA) is located in the nucleus of cells, while some are also found in the mitochondria (known as mtDNA). DNA is passed on to offspring. Therefore, it is used by people who want to trace their biological ancestors. The mtDNA in cells converts the energy from food into forms the cells use.

DNA letters: DNA stores its genetic information in a string of four chemicals: guanine (G), cytosine (C), adenine (A), and thymine (T). Geneticists use the letters to analyze genetic information and in genome projects to study the sequence of the DNA letters.

Evolutionary Creationism: This theory embraces the biblical idea of creation. However, it proposes that God *(Jesus)* created everything through a designed, orderly, gradual process of evolution over billions of years. It says that modern people evolved from pre-human ancestors over a lengthy period. During this time, the image of God and human sin gradually developed in people. Individuals who believe this refer to themselves as Christians because they love God and strive to live for him. They say they have a personal relationship with Jesus Christ and live by the power of the Holy Spirit. They say they experience the supernatural work of the Holy Spirit in their lives.

Ex Nihilo: *Ex Nihilo* is a Latin phrase that means "out of nothing." When applied to the creation account, it means Jesus created the universe and Earth out of nothing. The following verses help us to understand this: *God stretches the northern sky over empty space and hangs the earth on nothing* (Job 26:7 NLT).

Fact (in the Bible): This Scripture is indisputably clear in its meaning.

Faith: Biblical faith is informed faith. It is confidence and trust that God is who he says he is and will do what he says (Hebrews 11:1–3). Faith requires action, or it's false (James 2:26). Faith grows over time deep within the soul as people experience God and allow him to apply his living word to transform their hearts and lives. Belief in the theory of evolution is "blind faith" because there is no evidential substance to support it.

Fossil: Have you ever picked up a fossilized bone? If so, you noticed that it was much heavier than it would have been living. This is because the tissues were "mineralized." Over time, the tissues decayed and were replaced with minerals in the ground. Most of the time, only the hard tissues, such as bone and shells, survive long enough to become a fossil. Since bones and shells have pores, the pores are also filled with minerals. Thus, completely mineralized fossils are solid rocks.

Garden of Eden: God planted this beautiful garden (Genesis 2:8–15) as a perfect place for Adam and Eve to live and work (they were farmers). When Adam and Eve rebelled against God, they were cast out of the Garden. It disappeared from the Earth as a result.

Gene: Basic unit of genetic information that occupies a fixed position on a chromosome. It's composed of DNA.

Gene Pool: Collection of all the different genes within the population of a species.

Genesis 1:1 Gap Theory: People who support this theory believe there were two distinct initial creations described in Genesis chapter 1. First Initial Creation: They think the universe was made over a lengthy period (Genesis 1:1–2), which left the Earth formless and empty. Second Initial Creation: They believe the second creation starts with Genesis 1:3 and includes the remaining verses of Genesis describing the six days of creation. The gap is between the end of the first creation (Genesis 1:1–2) and the start of the second creation (Genesis 1:3). This theory has several variations.

Genesis Chapters One and Two Contradiction Theory: This theory about the need for a second initial creation is based on the supposition that Genesis chapters 1 and 2 contradict one another in their accounts of when plant life and Adam and Eve were created.

Note: God doesn't contradict himself in the Bible. Any suggested contradiction is the result of our not fully understanding God's methods and reasons described in the Bible.

Genetics: This science studies an organism's chromosomes, genes, DNA, RNA, and other genetic factors to determine traits and how and why the organism functions. Genetic information is used in many sciences to discover why an organism looks and behaves as it does.

Genus: The level above species in the biological classification of organisms. It's composed of a select number of species related to one another genetically.

God (Father God): When we see the word *God* in the Bible, it typically identifies the first person of the Trinity. Sometimes, it also refers to the Trinity, emphasizing their oneness. In the Hebrew language of the Old Testament, God gives himself one personal name (Exodus 3:13–14). Because the people of Israel thought of this as his sacred name, they referred to it as YHWH, leaving out the vowels. This is sometimes written in English Bibles as I AM WHO I AM, lord (with small capital letters), Adonai, or Jehovah. Many names and titles for God reveal aspects of his divine nature and represent who he is, such as *El Shaddai* (God Almighty) or *Jehovah-shalom* (God is our peace).

Heaven: When the physical body of a follower of Jesus Christ dies, their spirit/soul goes immediately to Heaven. The Father God, Jesus Christ, myriads of angels, and other believers welcome them with loving and joyful arms. Heaven is the eternal abode of God (Father God, his Son, Jesus Christ, and the Holy Spirit), even though they exist everywhere. Since God is supernatural, it's supernatural as well. It's the place where God presides over the entire creation and his multitude of mighty angels. Heaven above comes down to become the new Jerusalem at the end of time, where all believers will exist eternally, enjoying God's presence on the new earth (Revelation 21:1–3).

Hell: When the physical body of those who reject Jesus Christ dies, their spirit/soul goes immediately to Hell. After the Great White Throne Judgment, they will be sent to exist eternally in the Lake of Fire, where they will never experience God's presence, love, and peace (Revelation 20:11–15).

Heredity: Passing of traits (from genes) from parent to child.

Holy Spirit (Spirit of God and Christ): The Holy Spirit is identified as the third person of the Trinity. He has many names throughout the Bible, including the Spirit of God and Spirit of Christ. These indicate he is divine, fully God in nature, and part of the Trinity. The Holy Spirit opens people's hearts and minds to understand the Gospel that Jesus died for their sins, was buried, and resurrected to provide eternal life with God. When they profess their faith in Jesus Christ as Savior and ask him to be the Lord of their life, the Holy Spirit comes to live within them. Only then are they born-again and become a Christian.

Hominin: Evolutionists use the term "hominin" to represent modern humans (Homo sapiens), Neanderthals, and extinct species they consider to be human. These include the species of the genus Homo. Homo habilis is the most ancient of these purported human species.

Human (Human Being): Synonyms used in the book are people, mankind, and Homo sapiens. "Homo sapiens" is the scientific name for the human species.

Human Genome: All the genetic information for human beings.

Human Lineage Theory: The proposed species in evolution's typical human lineage theory are bipeds and hominins. Human lineage is based on direct descent.

Image of God: People are unique among all the creations orchestrated by our Intelligent Designer. According to the Bible, the Creator God made people in their image.

Intelligent Design: Everything created by Jesus was done based on intelligent designs. They did not just evolve without purpose, and purpose requires design. Modern genetics validates in many ways that intelligent design is behind the purpose and function of all cells.

Intelligent Designer (Father God): The Father God is the Intelligent Designer, and Jesus is the person of God who spoke creation into existence (Genesis chapter 1).

Jesus Christ (Savior and Lord): He is identified as the second person of the Trinity. Jesus is his personal name (Matthew 1:21). Christ is a title that means the one chosen by God to save his people. He is fully God and fully divine in his nature. He is found throughout the entire Bible and has many names and titles that help us understand who he is and what he does.

Lamarck's Theory of Evolution: French naturalist Jean Baptiste Pierre Antoine de Monet, Chevalier de Lamarck, in 1801, was the first to propose that species changed over long periods into new species. Prominent in his theory is "use" and "disuse" of organs by animals. He believed that as they adapted to their environments, nature caused them to evolve from simple to complex living forms.

Macroevolution: This is the theory that a long evolutionary process occurs by which species undergo significant genetic and trait changes and evolve into new species. This is essentially the theory of biological evolution of species. Science tells us that for macroevolution to occur, massive amounts of new genetic material must be created in a species for a new species to evolve from it. No examples of new genetic material in a species have been discovered to support this theory. So scientists substitute evolutionary mechanisms, such as gene mutations, genetic drift, gene migration, and natural selection, to theorize the concept of macroevolution.

Microevolution: This is the process of adaptation by which changes to a species' genetic makeup and physical traits occur in small ways over short periods. Many creationists accept this theory of environmental adaptation within a species. Some evolutionists theorize that microevolution has been the driving mechanism for the macroevolution process for billions of years. This, too, has never been proven scientifically to be true.

Millennial Reign of Christ: By the end of the seven-year Tribulation, the Earth will be completely destroyed. The Earth will not be able to sustain human or any other life. In Revelation, chapters 19 and 20, we read about the Millennial Reign of Christ on Earth that occurs just after this horrific period. We see that all the saved who were martyred by the antichrist and his armies will reign on Earth with Jesus for this thousand years.

Missing Link Theory: The theory of macroevolution requires the existence of "missing links." A missing link is a vacant placeholder to fill a gap in the evolutionary tree for a non-existent transitional species. Since they do not exist, there's a gap in the worldwide fossil records for a species. For example, for vertebrates to move out of the water onto land to walk, there must have been a transitional ancient sea creature that developed lungs to breathe air and feet to walk on land. This missing link and all others have never been discovered because they do not exist.

Modern Synthesis (Neo-Darwinian Theory): This is an updated version of Charles Darwin's Theory of Evolution by Natural Selection that embeds the study of genetics and modern population genetics. Modern Synthesis is evolutionary scientists' most commonly accepted theory of evolution today. However, some disagree in part or totally with its core principles.

Mutation: Mutations occur when a cell makes a mistake in copying its DNA and is unable to correct it. These uncorrected mistakes become part of the organism's DNA.

Mystery: There are some verses and topics in the Bible that are simply a mystery. A mystery in the Bible is something for which God has partially concealed the complete meaning or even hidden the entire meaning.

Myth: These occur everywhere in a society. However, there are no myths included in the Bible. For example, there are many myths about Heaven and Hell that are not in the Bible.

Natural Selection and Adaptation: The classical description of natural selection is that it's a process by which species (plants and animals) better suited to their environment survive and reproduce successfully due to random, non-directed genetic mutations.

Natural World: People live in a natural world designed by the Father God and created by Jesus for them to live within. The complexities and intricate designs of the natural world dictate there is an Intelligent Designer. The theory of evolution cannot explain these.

Neanderthal: Science has proven that this people group existed alongside modern humans and interbred with them. Many people today have 1 to 4 percent Neanderthal DNA due to this. We don't know for sure if they are a separate species of humans or the same species. We do know that they are not the proposed species predecessor to modern humans in the human lineage theory. As a creationist, I believe Jesus created them on Day Six, but that they do not have the image of God.

New Heaven and New Earth. God will ultimately terminate this current creation by fire because of the rampant rebellion of people against him. He has delayed doing so because he wants every person to repent and be saved. After its destruction, he will create an entirely new heaven and new Earth. This will be a wonderful place of eternal existence for the followers of Jesus Christ. The Father and Jesus Christ will come down from Heaven to live among them in this new creation. There will never again be pain, suffering, or sin in this eternal home with God. There will be no crime or murder because no one without Jesus Christ will be there.

Non-Random Gene Mutations: These mutations occur as cells sense their changing environment. In response, they create specific changes to genes that alter the organism's physical and behavioral traits to adapt successfully to the changing environment. These mutated genes sensed as harmful to the organism are typically destroyed within the cell. Some escape the cell's efficient monitoring and become part of the organism's genetic composition.

Prehistoric: Evolutionists say that anything that existed before the advent of writing is prehistoric.

Primordial Soup Theory: This is the theory that billions of years ago, the Earth had a "primordial soup" of inert chemicals necessary to produce living organisms. An initial theory was that lightning struck this soup to produce various amino acids, which are the "building blocks" of life. Once the essential amino acids were formed in the soup, they proposed life of some kind evolved over a very long period. Modern Science proves that life on planet Earth did not evolve from a primordial soup.

Radiometric Dating: Scientists, especially geologists, use radiometric dating to estimate the age of rocks and fossils contained within them. This is also called radioisotope dating or radioactive dating.

Random Gene Mutations: These gene mutations occur randomly, without direction from the cell. They are the result of errors within a cell as it copies DNA. Some are beneficial, and some are harmful. Those that are beneficial enable an organism to adapt to its environment successfully.

Redshift and Blueshift (lightwave frequencies): Edwin Hubble theorized their existence as a way to determine if a cosmic object was moving closer to or farther away from Earth. Astronomers use these as evidence to support the theory that the universe is expanding. They believe the universe's expansion stretches the wavelengths of light traveling through it. Redshift represents a longer (or stretched) light wavelength as objects move away from Earth. The blueshift represents a shorter light wavelength as objects move closer to Earth. Redshift and blueshift correspond to distances from Earth in light years. Some scientists today use this theory to provide evidence that the Big Bang Theory correctly describes the origin of the universe.

River of Life: The only freshwater mentioned in Revelation chapters 21 and 22 is the River of Life that flows from the throne of God in the New Jerusalem. This implies there is no need for fresh water for drinking and farming throughout the new Earth.

Sentient: This means people are capable of sensing what is occurring within and around them. They are conscious of and can thoughtfully respond to and interact with their environment. Self-awareness allows them to perceive what is happening in their minds and hearts. It means they have a soul that is beyond the physical world. Evolution can only attempt to explain the development of the natural and physical, not the soul and spirit.

Science: The sciences study the universe and the natural world created by Jesus. Whether they know it or not, scientists seeking truth without bias are learning about God because he is the Intelligent Designer behind creation. The Bible and science are not mutually exclusive. Instead, they support one another when reviewed from an objective perspective.

Scientific Fact: This is the result of repeated testing using the scientific method. The hypothesis has repeatedly been proven to be true under many different conditions. Therefore, objective observations and experimentations repeatedly verify it as a consistent fact.

Scientific Hypothesis: Hypotheses are observation-based. This is a proposed scientific explanation for something observed. It's an "educated guess" based on observation. Hypotheses are proved or disproved by scientific experimentation and the scientific method.

Scientific Method: Scientists seeking truth use the scientific method to prove their hypothesis is true. The scientific method is a four-step process involving observation, hypothesis, experimentation, and conclusions. If the hypothesis is proven valid and reliable, it is considered a scientific fact. However, if the hypothesis cannot always be proven true when tested, scientists may consider it a scientific theory that has some evidence of its validity.

Scientific Theory: Theories are evidence-based ideas but not proven facts. While theories are not irrefutable scientific facts, many scientists treat them as such.

Scientific Theory of Intelligent Design: This theory proposes that the complexity and intricate design of the universe and living organisms indicate an intelligent cause (design) rather than the random genetic variations of natural selection. It is also intended to show that intelligent design can occur naturally without an Intelligent Designer (as Christians believe).

Speciation: Evolutionists say that speciation occurs during the evolution of a plant or animal when a specific event interrupts an organism's normal life. Evolutionary scientists think significant geographical changes occur that separate the species into two or more geographic populations. As a result, organisms from these separated populations can no longer access one another for interbreeding. Over long periods, the populations better adapt to their environment through natural selection and develop genetic mutations that lead to genetically distinct species.

Species (singular and plural): These are groups of organisms related to one another in a specific scientific genus that can reproduce between themselves but not outside their genus. For example, human beings are classified in the genus "Homo" and species "sapiens" (Homo sapiens).

Supernatural World: There is something beyond the realm of the physical, natural world. It's the supernatural world. It's the unseen world around us. The Triune God and his angels are supernatural, and everything they do is supernatural. The devil and his demons are supernatural also, and everything they do is supernatural.

Theistic Evolution: The beliefs of Theistic Evolutionists are similar to Evolutionary Creationists in the following ways: God *(Jesus)* created everything, including the universe; Some of the theories of evolution are valid; and There was intelligent design in creation. The major difference is that Theistic Evolution focuses on the universe, while Evolutionary Creationism focuses on the biological evolution of species.

Tower of Babel: This account in the Bible represents adaptation by natural selection of human beings (microevolution). People have continued to adapt to their environments for thousands of years since the Tower of Babel (Genesis 11:1–9).

Trait: Physical or behavioral characteristics from genes (passed from parent to child).

Tree of Life (Biblical): The Tree of Life is mentioned in The Garden of Eden in Genesis and the book of Revelation on the new Earth. *On either side of the river was the tree of life, bearing twelve kinds of fruit, yielding its fruit every month* (Revelation 22:2 NASB). One significant mystery of the Tree of Life is why people will use its leaves.

Trinity (Triune God): The terms "Trinity" and "Triune God" don't appear in the Bible. Biblical scholars created them to help people understand the unity, yet separate reality, of the three persons of God. They are God the Father, his Son, Jesus Christ, and the Holy Spirit. They are each equally God in every aspect of their being and nature.

Truth: It's an absolute that doesn't change with the culture or social norms. No lies, deceptions, or half-truths are involved with biblical truth. Everything in the Bible is true because it's *God-breathed* (2 Timothy 3:16).

Unidentified Flying Object (UFO): This is a term used to label any object seen in our global skies that could not be clearly identified. Unidentified Anomalous Phenomena (UAP) has replaced the UFO name for unidentified objects in an attempt to remove the conspiracy stigma that might prevent reportings of purported sightings to the government. They want to know what people believe they have witnessed so they can investigate them. Identified Flying Objects (IFOs) are UFOs that have been identified.

Universe: In Genesis 1:1, we see the creation of the first universe, and in Revelation 21:1, we see the uniquely created new universe.

World: The term world in the context of many scriptures (such as John 16:7–11) doesn't mean the physical planet Earth. Instead, it refers to a society of people without God that has its own morals, standards, and values.

World Religions: All major world religions (except Buddhism) believe in the existence of one or more gods. You can learn more details about their beliefs by searching the Internet.

# Bibliography

Active Wild Admin. "List Of Dinosaurs: Dinosaur Names With Pictures & Interesting Information." Accessed 12/14/2024. https://www.activewild.com/list-of-dinosaurs-names-with-pictures/#dinosaur-index.

All About The Journey. "Human Eye." Accessed 1/14/2025. https://www.allaboutthejourney.org/human-eye.htm.

Allen, Cecil. "A Fossil Is a Fossil Is a Fossil. Right?" Accessed 12/19/2024. https://creation.com/a-fossil-is-a-fossil-is-a-fossil-right.

American Oceans. "Water Dinosaurs." Accessed 12/15/2024. https://www.americanoceans.org/facts/water-dinosaurs.

Amos, Jonathan. Science Correspondent, BBC News. "Amazing Images from James Webb Telescope, Two Years after Launch." Accessed 4/5/2024. https://www.bbc.co.uk/news/resources/idt-611525eb-3a0c-4a68-bf54–485df138b6f6.

Ancient Code Team. "5 Ancient Petroglyphs & Cave Paintings that Depict 'Ancient Aliens.'" Accessed 11/28/2024. https://www.ancient-code.com/5-ancient-petroglyphs-cave-paintings-that-depict-ancient-aliens.

Andrews, Peter. "Hominoid Evolution." *Nature* Vol. 295, No. 5846 (1982).

Answers in Genesis. "Age of the Universe." Accessed 2/21/2024. https://answersingenesis.org/astronomy/age-of-the-universe.

Barna, George, PhD. Director of Research, Cultural Research Center at Arizona Christian University. "American Worldview Inventory 2023, Release #1: Incidence of Biblical Worldview Shows Significant Change, Since the Start of the Pandemic." Accessed 1/13/2025. https://www.arizonachristian.edu/wp-content/uploads/2023/02/CRC_AWVI2023_Release1.pdf.

Barnes, Albert. "Albert Barnes Notes on the Bible." Accessed 3/24/2024. https://www.e-sword.net. Philadelphia: Public domain, 1834.

BD Editors. "Speciation." Accessed 6/3/2024. https://biologydictionary.net/speciation.

Bechly, Günter. "Fossil Friday: New Evidence for the Human Nature of Neanderthals." Accessed 6/4/2024. https://evolutionnews.org/2024/02/fossil-friday-new-evidence-for-the-human-nature-of-Neanderthals.

Berkeley University. "What is macroevolution?" Accessed 1/14/2025. https://evolution.berkeley.edu/evolution-101/macroevolution/what-is-macroevolution.

Biblical Archaeology Society. *Biblical Archaeological Review Magazine*. https://www.biblicalarchaeology.org/biblical-archaeology-review.

Blaxland, Beth, and Fran Dorey. "Walking on Two Legs—Bipedalism." Accessed 3/16/2024. https://australian.museum/learn/science/human-evolution/walking-on-two-legs-bipedalism.

Bowden, Malcolm. *Ape-Men: Fact or Fallacy*. Bromley, Kent, England: Sovereign, 1977.

Bowling, Trey with Dr. Brian Thomas, "The Soulless Hominid Theory: A Fatal Flaw in Old Earth Creationism." Accessed 3/30/2025. https://www.icr.org/article/15206.

Bryan, Paul. *Who Is This God? A Handbook for Life with Him*. Houston: Lucid, 2023.

Chamary, JV. "Does the tree of life reflect evolution? Does evolution's 'tree of life' accurately reflect the relationships between everything that has ever lived?" Accessed 4/13/2024. https://www.discoverwildlife.com/animal-facts/tree-of-life-evolution.

Clarey, Dr. Tim. *The Science of the Biblical Account, Dinosaurs, Marvels of God's Design*. Green Forest: Master Books, 2015, 2022.

Clarke, Adam. "Adam Clarke's Commentary of the Bible." Accessed 3/24/2024. https://www.e-sword.net. Britain: Public domain, 1831.

Coffey, Donavyn. "How Does DNA Know Which Job to Do in Each Cell?" Accessed 1/11/2025. https://www.livescience.com/how-dna-turns-on-off.html.

Cornell University. "Lingering effects of Neanderthal DNA found in modern humans." Accessed 1/13/2025. https://news.cornell.edu/stories/2023/06/lingering-effects-neanderthal-dna-found-modern-humans.

Crockett, Christopher. "What Do Redshifts Tell Astronomers?" Accessed 1/16/2025. https://earthsky.org/astronomy-essentials/what-is-a-redshift.

Dinosaurs Facts for Kids. "List of Flying Dinosaur Names: All Pterosaur Species!" Accessed 12/14/2024. https://dinosaurfactsforkids.com/list-of-flying-dinosaur-names-all-pterosaur-species.

Faulkner, Danny R. and Lee Anderson Jr. *The Created Cosmos, What the Bible Reveals About Astronomy*. Green Forest: Master Books, 2016.

Garris, Zachary. "What Do The Genealogies Of Genesis 5 & 11 Teach About The Age Of The Earth?" Accessed 6/9/2024. https://knowingscripture.com/articles/what-do-the-genealogies-of-genesis-5-and-11-teach-about-the-age-of-the-earth.

Genesis Park. Genesis Park, "Fossil Footprints." Accessed 12/9/2024. https://www.genesispark.com/exhibits/evidence/paleontological/footprints.

Gill, John. "John Gill's Expository of the Entire Bible." Accessed 3/24/2024. https://www.e-sword.net. England: Public domain, 1746 to 1763.

Grandchamp, Greg. "What Is Young Earth Creationism?" Accessed 1/14/2025. https://www.christianity.com/wiki/bible/what-is-young-earth-creationism.html.

Greene, Jon W. "A Biblical Case for Old-Earth Creationism." From *Godandscience.org*. Accessed 1/14/2025. http://godandscience.org/youngearth/old_earth_creationism.html https://www.scienceandfaith.org/old-earth-creationism.

Gregory, William K. "*Hesperopithecus* Apparently Not an Ape nor a Man." *Science* Vol. 66, No. 1720 (December 16, 1927).

Haarsma, Deborah B. *Four Views on Creation, Evolution, and Intelligent Design*. Grand Rapids: Zondervan Academic, 2007.

Ham, Ken. "Answers in Genesis." Accessed 2/21/2024. https://answersingenesis.org.

———. "A Mature Universe." Accessed 4/5/2024. https://answersingenesis.org/astronomy/age-of-the-universe/mature-for-her-age.

Hashimoto, Todd. "How Much Water In Lake Superior?" Accessed 3/9/2024. https://www.lakebeyond.com/how-much-water-in-lake-superior.

Heguy, Adriana. Accessed 3/16/2024. https://www.quora.com/Is-there-DNA-evidence-that-Homo-erectus-was-a-different-species-from-Homo-sapiens.

Helmenstine, Anne. "Elements in the Human Body and What They Do." Accessed 6/12/2024. https://sciencenotes.org/elements-in-the-human-body-and-what-they-do.

Helmenstine, Todd. "Periodic Table Wallpaper With All 118 Elements." Accessed 3/9/2024. https://sciencenotes.org/periodic-table-wallpaper-118-elements.

Hey, Jody. PhD. "Why Should We Care about Species?" Accessed 5/3/2024. https://www.nature.com/scitable/topicpage/why-should-we-care-about-species-4277923.

IntelligentDesign.org. "What Is Intelligent Design?" Accessed 6/5/2024. https://intelligentdesign.org/whatisid.

Joy Imes, Carmen. *Being God's Image, Why Creation Still Matters*. Downers Grove: Intervarsity Press, 2023.

Kennedy, Titus. *Unearthing the Bible, 101 Archaeological Discoveries that Bring the Bible to Life*. Eugene: Harvest House, 2020.

———. *Excavating the Evidence for Jesus: The Archaeology and History of Christ and the Gospels*. Eugene: Harvest House, 2022.

Kindy, David. "Largest Human Family Tree Identifies Nearly 27 Million Ancestors." Accessed 6/11/2024. https://www.smithsonianmag.com/smart-news/largest-human-genomic-family-tree-identifies-nealy-27-million-ancestors-180979657.

Klink III, Edward W. *The Beginning and End of All Things, A Biblical Theology of Creation and New Creation*. Downers Grove: Intervarsity Press, 2023.

Lamoureux, Denis O. "Evolutionary Creation: A Christian Approach to Evolution" Accessed 5/20/2024. https://www.scienceandfaith.org/evolutionary-creationism.

Lea, Robert. "NASA Exoplanet Telescope Discovers 'Super-Earth' in Its Star's Goldilocks Zone. TESS Saw the Planet May also Have an Earth-sized Companion." Accessed 4/1/2024. https://www.space.com/exoplanet-super-earth-habitable-zone-tess.

Lester, Lane. "Genetics: No Friend of Evolution." Accessed 11/2/2024. https://creation.com/genetics-no-friend-of-evolution.

Loring, Brace C, and Ashley Montegu. *Human Evolution: An Introduction to Biological Anthropology*. New York: MacMillan, 1977.

Lund University. "AI-based Method for Dating Archaeological Remains." *ScienceDaily*, 23 August 2022. https://www.sciencedaily.com/releases/2022/08/220823162730.htm.

May, Andrew. "What is the Big Bang Theory?" Accessed 4/5/2024. https://www.space.com/25126-big-bang-theory.html.

Mehlert, Albert W. "Homo Habilis Dethroned." *Contrast: The Creation-Evolution Controversy* Vol. 6, No. 6. Minneapolis: November/December, 1987.

———. "Lucy—Evolution's Solitary Claim for an Ape/Man: Her Position Is Slipping Away." *Creation Research Society Quarterly* Vol. 22, No. 3 (December 1985).

Metaxas, Eric. *Is Atheism Dead?* Washington D.C: Salem Books, 2021.

Miller, Mark. "Ancient Babylonian Tablet Provides Compelling Evidence that the Tower of Babel Did Exist." Accessed 5/31/2024. https://www.ancient-origins.net/news-history-archaeology/ancient-babylonian-tablet-provides-compelling-evidence-tower-babel-did-021378.

Morris, Henry M. "The Ice Age." Accessed 3/9/2024. https://creation.com/the-ice-age.

Naselli, Andrew David, and J.D. Crowley. *Conscience. What It Is. How to Train It. And Loving Those Who Differ*. Wheaton: Crossway, 2016.

National Aeronautics and Space Administration (NASA). "Mysteries of the Universe." Accessed 4/5/2024. https://www.nasa.gov/specials/60counting/universe.html.

National Aeronautics and Space Administration (NASA) Science, SpacePlace, Explore Earth and Space. "What is dark matter?" Accessed 1/12/2025. https://spaceplace.nasa.gov/dark-matter/en.

National Archives. "America's Founding Documents. Declaration of Independence Transcript." Accessed 1/25/2025. https://www.archives.gov/founding-docs/declaration-transcript.

National Human Genome Research Institute. "Human Genome Project Results." Accessed 5/31/2024. https://www.genome.gov/human-genome-project/results.

———. "Chromosomes Fact Sheet." Accessed 5/31/2024. https://www.genome.gov/about-genomics/fact-sheets/Chromosomes-Fact-Sheet.

National Oceanic and Atmosphere Administration (NOAA). "The Atmosphere." Accessed 11/28/2024. https://www.noaa.gov/jetstream/atmosphere.

Neyman, Greg. "Word Study—Yom." Accessed 2/25/2024. https://www.oldearth.org/word_study_yom.htm.

The Nine Planets, "What would happen if there was no moon?" Accessed 1/14/2025. https://nineplanets.org/questions/what-would-happen-if-there-was-no-moon/

Obreschkow, Dr Danail. "Cosmic Eye 'Louise.' A Zoom Journey from the Universe to the Sub atomic." Accessed 4/1/2024. https://www.europa.uk.com/cosmic-eye-louise.

O'Connell, Patrick. *The Science of Today and the Problems of Genesis*. Hawthorne, California: Christ Book Club, 1969.

Oxnard, Dr. Charles. *Fossils, Teeth, and Sex: New Perspectives on Human Evolution*. Seattle: University of Washington Press, 1987.

———. *The Order of Man: A Biomathematical Anatomy of Primates*. New Haven: Yale University Press, 1984.

Parker, Ray, Reviewer. "The Staggering Complexity of the Human Brain. Why our brains are the most complex structures in the known universe." *Psychology Today, Neuroscience*. Accessed 1/14/2025. https://www.psychologytoday.com/us/blog/consciousness-and-beyond/202309/the-staggering-complexity-of-the-human-brain.

The Planets. "Moons in the Solar System." Accessed 1/16/2025. https://theplanets.org/moons.

Prehistoric Saurus. "How Do Fossils Form: From Life to Stone." Accessed 12/19/2024. https://prehistoricsaurus.com/dinosaur-fossils/fossilization-process/how-do-fossils-form.

Price, Corey. "The Most Abundant Elements In The Earth's Crust." Accessed 11/10/2024. https://www.worldatlas.com/environment/the-most-abundant-elements-in-the-earth-s-crust.html.

Price, Michael. "Study reveals culprit behind Piltdown Man, one of science's most famous hoaxes" *Science Magazine*. Accessed 1/14/2025. https://www.science.org/content/article/study-reveals-culprit-behind-piltdown-man-one-science-s-most-famous-hoaxes.

Pruitt, Sarah. "What Happened at the Scopes Trial?" *History Channel*. Accessed 1/1/2025. https://www.history.com/news/90-years-ago-scopes-and-evolution-indicted-in-tennessee.

Pultarova, Tereza, and Daisy Dobrijevic. Contributions from Elizabeth Howell and Nola Taylor Tillman. "Milky Way Galaxy: Everything You Need to Know about Our Cosmic Neighborhood." Accessed 5/22/2024. https://www.space.com/19915-milky-way-galaxy.html.

Rensberger, Boyce, and Jay Matternes. "Facing the Past," *Science* Vol. 2 No. 8 (October 1981).

Ritchie, Hannah. "How Many Species Are There?" Accessed 3/9/2024. https://ourworldindata.org/how-many-species-are-there.

Robinson, Anthony. "How Many Planets are in the Universe? (A Staggering 22 Sextillion!)" Accessed 1/15/2025. https://skiesandscopes.com/how-many-planets.

Roser, Max, and Hannah Ritchie. "The World Population Has Increased Rapidly in Recent Centuries. But This Is Slowing." Accessed 3/16/2024. https://ourworldindata.org/population-growth-over-time.

Sapiens. "Hominins." Accessed 3/16/2024. https://www.sapiens.org/teaching-unit/hominins.

Schwitzgebel, Eric, and Jacob Barandes. "Could the Universe Be Finite? It's Not Absurd to Think the Universe Might Endure Forever." Accessed 4/1/2024. https://nautil.us/could-the-universe-be-finite-466593.

Science.nasa.gov. "Star Types." Accessed 4/6/2025. https://science.nasa.gov/universe/stars/types.

———. "What Is the Universe?" Accessed 4/6/2025. https://science.nasa.gov/exoplanets/what-is-the-universe.

Sci News Staff. "Some Plesiosaurs May Have Lived in Freshwater River Systems." *SciNews.com*. Accessed 12/5/2024. https://www.sci.news/paleontology/freshwater-plesiosaurs-11047.html.

Siegel, Ethan. Senior Contributor. "This Is Where The 10 Most Common Elements In The Universe Come From." Accessed 11/10/2024. https://www.forbes.com/sites/startswithabang/2020/05/25/this-is-where-the-10-most-common-elements-in-the-universe-come-from.

Smithsonian Institution. "Ancient DNA and Neanderthals." Accessed 5/31/2024. https://humanorigins.si.edu/evidence/genetics/ancient-dna-and-Neanderthals.

Stern, Jack T., and Randall L. Sussman. "The locomotor anatomy of Australopithecus afarensis." *Journal of Physical Anthropology*. Vol. 60, Issue. 3. (March, 1983).

Sulloway, Frank J. "The Evolution of Charles Darwin." Accessed 4/13/2024. https://www.smithsonianmag.com/science-nature/the-evolution-of-charles-darwin-110234034.

Tackett, Dr. Del. "Is Genesis History?" Accessed 2/21/2024. https://isgenesishistory.com.

Taylor, Paul S. *The Illustrated Origins Answer Book, Concise, Easy-to-Understand Facts about the Origin of Life, Man, and the Cosmos*. 4th ed. Mesa: Eden, 1992.

Team Biology Simple. "Cladogram." Biology Simple, March 1, 2024. https://biologysimple.com/cladogram.

Than, Ker. "God Particle Found? Historic Milestone. From Higgs Boson Hunters." Accessed 6/22/2024. https://www.nationalgeographic.com/science/article/120704-god-particle-higgs-boson-new-cern-science.

Thomas, Brian PhD. "DNA in Dinosaur Bones?" Accessed 1/14/2025. https://www.icr.org/article/dna-dinosaur-bones.

Tomkins, Jeffrey P. PhD. "The Untold Story Behind DNA Similarity." *Answers Magazine*. Accessed 9/12/2024. https://answersingenesis.org/genetics/dna-similarities/untold-story-behind-dna-similarity.

———. "Engineered Parallel Gene Codes Defy Evolution." Accessed 5/31/2024. https://www.icr.org/article/14831.

Walker, Tas. "Where Did All the Water Go?" Accessed 3/9/2024. https://creation.com/where-did-all-the-water-go.

Wellman, Jared. "Does the Genesis Creation Account Come from the Babylonian Enuma Elish?" Accessed 2/28/2024. https://carm.org/other-questions/does-the-genesis-creation-account-come-from-the-babylonian-enuma-elish.

Wetzel, Ralph M., et al. "*Catagonus*, An 'Extinct' Pecarry, Alive in Paraguay," *Science*, Vol. 189, No. 4200 (August 1, 1975).

Wikipedia. "Y-chromosomal Adam." Accessed 3/16/2024. https://en.wikipedia.org/wiki/Y-chromosomal_Adam.

Wood, Todd. "How Many Human Species?" *(blog, June 26, 2020)*. https://newcreation.blog/how-many-human-species.

Zhao, Buyun. "Charles Darwin & Evolution 1809–2009. The Modern Synthesis" Accessed 4/13/2024. http://darwin200.christs.cam.ac.uk/modern-synthesis.

Zondervan. *Archeology Study Bible, An Illustrated Walk Through Biblical History and Culture.* Grand Rapids: Zondervan, 2005.

# Index

Adam and Eve, 30–32, 35, 47–53, 84, 92, 99, 121, 123–25, 130, 135, 149–51, 164–65, 186, 211, 213–14, 221, 236, 240, 251, 270, 294
Adapt, 3, 10, 84, 91, 95–96, 98, 120–21, 165, 173–74, 176, 181–83, 205, 207, 212, 214–16, 219, 293, 296–99
Adaptation, 3, 121, 165, 174, 176, 182–83, 212, 215–16, 296–97, 299
Adapted, 98, 174, 181, 219, 296
Adapts, 212
Alien, 235–41, 291, 301
Aliens, 235–41, 291, 301
Ancestor, 38, 62–63, 67, 98, 102, 113, 115–17, 120, 164, 173, 175–76, 186, 213–14, 221, 278, 291, 293, 302
Ancestors, 38, 62–63, 67, 98, 102, 113, 115, 117, 120, 164, 173, 186, 213–14, 278, 293, 302
Ancient, 6, 38, 62–63, 65, 69, 97–98, 102, 113, 115–17, 120, 122, 165, 173–76, 193, 196, 200, 213–14, 221, 231, 238–39, 243, 278, 291, 295–96, 301, 303–4
Archaeologist, 193 95, 197, 200, 239
Archaeologists, 193–95, 200, 239
Archaeology, 38, 193, 195, 197–98, 200, 291, 301–2
Atmosphere, 82–83, 85, 105–6, 187–88, 236–37, 253, 255–57, 261, 291, 303

Biological, 4, 52, 62, 115, 124, 171, 173–74, 177, 182–84, 187, 189, 201, 208–10, 212–14, 219–22, 273, 278, 282, 293–94, 296, 299, 303
Biologist, 200, 215
Biologists, 200, 215
Biology, 84, 100, 115–16, 172, 182, 187, 189, 200–201, 221, 291, 301, 304
Biology's Tree of Life, 187, 291
Bipedalism, 62, 114, 291, 301
Blueshift, 244, 247, 298

Cell, 3, 85, 96, 109, 185, 188, 204, 209–17, 220, 291, 293, 295–98, 302
Cells, 3, 85, 96, 109, 185, 204, 210–12, 214–16, 220, 291, 293, 295, 297
Charles Darwin, 10, 173–74, 182, 187, 292, 296, 304–5
Chromosome, 200, 209–11, 214, 220, 222, 291, 294, 303
Chromosomes, 200, 209–11, 214, 291, 294, 303
Cladogram, 221, 291, 304
Cladograms, 221, 291
Conscience, 131, 137, 205–8, 292, 303
Cosmologist, 229, 232, 292
Cosmologists, 229, 232, 292
Cosmology, 244, 292
Cosmos, 116, 239–40, 244, 292, 302, 304
Creationism, 35–37, 51–53, 100, 126, 175, 177, 186, 189, 292–93, 299, 301–2
Creationist, 35, 38–39, 51–53, 55, 57, 59, 62–63, 98–103, 105–8, 110, 113–15, 119, 183–84, 187, 200, 212–14, 219, 226–29, 232, 246–47, 292, 296–97, 299
Creationists, 35, 38–39, 51–53, 55, 57, 59, 62, 98–103, 105–8, 110, 113–15, 119, 183–84, 187, 200, 212–14, 219, 226–29, 232, 246, 292, 296, 299
Creation Kind, 37, 78–79, 84–85, 87, 91, 95–96, 119, 123, 125, 176, 219–22, 241, 282, 285, 292
Creation Kinds, 78–79, 84, 87, 95–96, 119, 125, 219–22, 241
Creator, 3–4, 6, 15–17, 20, 35–39, 41–42, 44, 67–68, 128–30, 133, 137, 149, 151, 153, 166–67, 205–6, 213, 215, 227, 230, 232, 235, 243, 246–47, 278, 281, 283, 285

Dark Energy, 78, 229–30, 232, 256, 292
Dark Matter, 78, 229–30, 232, 254–56, 292, 303

Design, 3, 17, 21, 36, 39, 42, 44, 51, 67, 75–76, 78, 80, 82–85, 91–92, 94–96, 100, 103, 129, 131–32, 161, 163–67, 174, 185–87, 189, 199, 201–12, 214–17, 220, 226–27, 229–31, 235–36, 238, 240, 243, 246, 278, 293, 295, 297–99, 301–2

Designer, 3, 17, 36, 39, 67, 185, 199, 201, 203–4, 212, 214, 295, 297–99

Designs, 78, 84, 96, 103, 185, 203, 214, 295, 297

Diversity, 28, 90, 92, 96, 103, 164, 219–21

DNA, 78–79, 109–11, 114–17, 119–22, 125–26, 209–10, 212–15, 217, 221, 278, 281, 291, 293–94, 296–98, 302, 304

Environment, 10, 42, 44, 59, 84, 90–91, 108, 110–11, 120, 125, 165, 173–74, 176, 181–83, 200, 212, 214–15, 219, 236–37, 271, 293, 296–99, 303

Environmental, 84, 110–11, 183, 212, 215, 219, 296

Environments, 44, 90, 108, 120, 165, 181, 183, 219, 236–37, 296, 299

Ex Nihilo, 36, 75, 134, 188, 228, 260, 293

Extinction, 100, 105–7, 109, 111, 124

Fact, 3, 5–7, 9–12, 36–37, 42, 56, 83, 91, 95, 97–98, 116, 125–26, 128, 165, 172, 176, 181, 197, 201, 209–10, 230–31, 237, 245, 279, 293–94, 298–99, 301–4

Facts, 3, 5–6, 9–12, 36–37, 83, 97, 116, 165, 172, 197, 201, 230, 237, 279, 299, 301–2, 304

Fossil, 3, 97, 100–101, 105, 107–11, 114–17, 119–22, 173, 175–76, 293–94, 296–97, 301–3

Fossilization, 107–8, 303

Fossilized, 97, 107–8, 110, 120, 293

Fossils, 3, 97, 101, 105, 107–9, 111, 114–15, 176, 294, 297, 303

Gap Theory, 49, 294

Garden, 30–32, 35–36, 47, 163–64, 174, 186, 213, 240, 270, 294, 299

Gas, 76, 82, 105–6, 228, 231, 236–38, 254–55

Gasses, 105, 236–38, 255

Gene, 3, 6, 15–16, 18–20, 25, 29–30, 32, 35–39, 42–43, 47–51, 53, 55–57, 59, 61–63, 65–68, 77–79, 81–87, 89, 92–93, 95–96, 98, 100, 102, 106, 109, 113, 115–17, 119–25, 127, 129, 132, 136, 141–42, 151, 153–55, 157–58, 161–65, 173–76, 182–83, 186, 196, 200–201, 203, 205, 209–17, 219–22, 225, 228, 232, 236, 241, 246, 251, 254, 260–62, 270, 278, 282, 284, 291–304

Genes, 15–16, 18–20, 29–30, 32, 35–39, 42–43, 47–51, 53, 55–57, 59, 61–62, 65–68, 77, 81–87, 89, 92–93, 95, 98, 102, 106, 113, 115–16, 120, 123, 127, 129, 132, 136, 142, 151, 153, 155, 158, 161, 163–65, 174, 176, 182–83, 186, 196, 200–201, 209–15, 219–21, 225, 228, 232, 246, 251, 260–62, 270, 278, 282, 284, 292, 294–95, 297, 299–304

Genetics, 3, 38, 79, 85, 116–17, 119–20, 174, 182, 200, 209–21, 294–96, 303–4

Genus, 100, 114, 173, 213, 220–21, 292, 294–95, 299

Heavens, 5, 11, 16, 23, 31, 41–42, 47–48, 53, 57, 59, 68–69, 81, 83, 86, 89, 95, 107, 132, 136, 141, 153–54, 161, 163, 225, 227–28, 232, 245, 249, 251–57, 260–72, 292

Hominin, 113–14, 122, 295, 304

Hubble, 230, 244, 298

Human Lineage, 114–15, 117, 122, 162, 176, 278, 295, 297

Hypotheses, 9, 298

Hypothesis, 9–10, 298

Hypothetical, 113, 221, 235, 291

Ice Age, 100, 106–7, 119, 303

Image, 42, 48, 50–52, 66, 113, 123–39, 149–52, 154, 161, 186, 206, 230–31, 239–40, 254, 266, 292–93, 295, 297, 301–2

Intelligence, 85, 185, 205, 214–15

Intelligent, 3, 17, 21, 36, 39, 67, 75, 78, 96, 99, 129, 163, 185, 187, 189, 199, 201, 203–4, 206, 208, 212, 214–17, 231, 235–36, 238, 243, 278, 291, 295, 297–99, 302

Invisible, 27–28, 129, 138, 141, 143–44, 161, 227, 229, 292

Lamarck, 181, 184, 296

Macroevolution, 101, 173–75, 182–83, 212–13, 215–16, 219, 296, 301

Mendel, 212

Meteor, 105–6

Microevolution, 121, 165, 174, 176, 183–84, 212, 219, 296, 299

Millennial, 252, 296

Mineral, 77, 90, 107–9, 120, 133, 293–94

Mineralization, 108

Minerals, 77, 90, 107, 133, 293–94

Missing Link, 114–16, 122, 175–77, 221, 296

Missing Links, 114–15, 175–77, 221, 296

Mutation, 3, 124, 174, 182, 210, 212–17, 219, 296–99

Mutations, 3, 124, 174, 182, 213–17, 219, 296–99

Myths, 3, 5–7, 11, 65, 67, 69, 71, 194, 279, 297

Natural Selection, 3, 10, 173–74, 181–82, 185, 212, 214, 219, 292–93, 296–97, 299
Natural World, 3–4, 7, 9–10, 37, 128, 137, 149, 161, 173–74, 177, 185, 201–2, 205, 230, 297–99
Neanderthal, 3, 84, 113–16, 119–27, 130, 278, 281, 295, 297, 301–2, 304
Neanderthals, 3, 84, 113–16, 119–27, 130, 281, 295, 301, 304
New Earth, 20, 47–48, 53, 141–45, 153, 157, 174, 249, 253–57, 259–73, 297–99
New Heaven, 48, 141–42, 249, 251, 260, 297
New Heavens, 47–48, 53, 141, 153, 245, 249, 251–57, 260–72
Non-Random, 3, 215, 217, 297

Offspring, 123
Old Earth Creationism, 37, 51–53, 126, 175, 177, 301
Old Earth Creationist, 51, 53, 57, 62–63
Old Earth Creationists, 35, 51–53, 55, 292
Organism, 79, 171, 173, 181–82, 185, 187, 199–200, 204, 209, 212–13, 215–16, 220, 291, 293–94, 296–99
Organisms, 79, 171, 173, 181–82, 185, 187, 199–200, 212, 220, 291, 293–94, 297, 299
Oxygen, 77–78, 91, 107–8, 132, 211, 236–37, 252, 255, 262

Periodic Table, 76, 78, 80, 132, 163, 236, 253, 260, 264, 302
Population, 62–64, 67–68, 106, 109, 164, 173, 182–83, 210, 294, 296, 299, 304
Populations, 67, 182, 299
Prehistoric, 97, 99, 108, 193, 200, 291, 297, 303
Primordial, 187–89, 297

Radiometric, 109, 297
Redshift, 244, 247, 298, 302

Science, 3–12, 38–39, 52, 61, 63, 76, 79, 100, 109, 114–15, 117, 120–21, 125, 151, 172, 174, 186, 188, 191, 193–217, 220, 222, 226–27, 229–30, 236–37, 240–41, 244, 281–82, 291, 293–94, 296–98, 301–4
Sciences, 3–4, 6–7, 9, 12, 39, 172, 199–202, 209, 229, 281–82, 294, 298
Scientific, 9–12, 38, 42, 51, 76, 98, 100–102, 115–16, 120, 125, 127–28, 131, 173–76, 181, 183, 185, 189, 197, 199, 201–2, 213–15, 229–30, 235–37, 240, 244–45, 256, 295–96, 298–99
Sentience, 125

Sentient, 125, 173, 205–6, 208, 235–36, 238, 291, 298
Skies, 225, 238, 300, 304
Sky, 10, 41, 43, 50, 69, 75, 77, 81–83, 85, 91–92, 95–99, 129, 132, 151, 183, 220, 226–27, 253, 261, 293
Species, 4, 10, 37, 52, 79, 84–85, 90–91, 93, 96–97, 99–101, 108, 110, 113–14, 116, 119, 121–23, 125, 127, 165, 171, 173–77, 181–84, 187, 200, 210, 212–15, 219–22, 278, 291–97, 299, 302, 304–5
Survival, 42, 124–25, 173, 181–82, 215
Survive, 90, 107, 124, 173–74, 181, 183, 212, 214–15, 255, 278, 293, 297
Survives, 90

Theistic, 187, 189, 299
Theoretical, 77, 124, 174, 187, 221, 228, 230, 236–37, 240, 243, 245, 291–92
Trinity, 15–17, 21, 23, 38, 50, 67–68, 135, 139, 141, 277–78, 289, 294–95, 299
Triune God, 3, 15–16, 21, 23, 28, 41, 66, 68, 135–37, 139, 230, 240, 299

UFO, 300
Unidentified Flying Object, 238–40, 300
Universe, 4, 9–10, 19, 23, 27, 35–39, 42, 44, 48–49, 51–52, 61–62, 64, 67, 69, 75–76, 78, 80–82, 85–86, 116, 136, 174, 185, 187, 199–200, 204, 223, 225–33, 235–36, 238–40, 243–73, 292–94, 298–304

Visible, 11, 27–28, 78, 85, 109, 129, 138, 142–46, 150, 161, 229, 232, 239, 246, 292
Volcanic, 106
Volcano, 105, 182

Water of Life, 272
Word of God, 61, 82, 127, 171, 194
World Religion, 66–71, 300

Young Earth Creationism, 37, 51, 53, 100, 302
Young Earth Creationist, 51, 53, 62–63, 100, 246–47
Young Earth Creationists, 35, 38–39, 51, 53, 55, 57, 59, 62, 107–8, 228, 292

www.ingramcontent.com/pod-product-compliance
Lightning Source LLC
Chambersburg PA
CBHW081144230426
43664CB00018B/2801